SpringerBriefs in Electrical and Computer Engineering

Jiannong Cao · Xuefeng Liu

Wireless Sensor Networks for Structural Health Monitoring

 Springer

Jiannong Cao
Hong Kong Polytechnic University
Kowloon
Hong Kong

Xuefeng Liu
Hong Kong Polytechnic University
Kowloon
Hong Kong

ISSN 2191-8112   ISSN 2191-8120 (electronic)
SpringerBriefs in Electrical and Computer Engineering
ISBN 978-3-319-29032-4   ISBN 978-3-319-29034-8 (eBook)
DOI 10.1007/978-3-319-29034-8

Library of Congress Control Number: 2016930052

Printed on acid-free paper

This Springer imprint is published by SpringerNature
The registered company is Springer International Publishing AG Switzerland

Preface

The Motivation for This Book

The past decade has witnessed the emergence of applying wireless sensor networks (WSNs) to monitor the healthy condition of civil infrastructures. Compared to conventional wire-based structural healthy monitoring (SHM) systems, a WSN-based SHM system has the advantages of low cost, ease of deployment, and can obtain fine-grained information about structure's condition.

The main motivation for offering this book stems from the observation that, although many WSN-based SHM systems have been deployed, some real requirements in SHM have not been fully addressed. SHM applications have some distinct features from traditional WSN applications like environmental monitoring. Thus many widely adopted techniques in conventional applications of WSNs, like the event-triggered wake-up, in-network processing, fault-tolerance, cannot be directly applied for SHM applications. In this book, we first give a review of existing WSN-based SHM systems, and then introduce our WSN-based platform called SenetSHM. SenetSHM adopts many techniques that are specially designed to address the unique features of SHM applications. We share our experiences by stepping from the hardware and software design, to in-field experiments of the SenetSHM.

What This Book Is About

This book provides comprehensive coverage and detailed insights into the emerging area of using WSNs for SHM. It helps the readers to understand the specific requirements of SHM applications from other traditional WSN applications, and how these requirements are addressed using a series of systematic approaches. Therefore, it can be seen as a textbook as well as a practical guide for the reader:

- To understand the state-of-the-art technologies in domain-specific applications of WSNs like SHM.
- To learn about the methodologies of how to address the specific requirements for a WSN application. In particular, we provide a guideline for problem formulation, problem solving, and share our experiences and lessons learned from our practices in implementing the SenetSHM.

How This Book Is Organized

This book is divided into seven chapters.

Chapter 1: Introduction. In this chapter, we introduce the background of SHM, some of the existing SHM systems, followed by the challenges and issues in developing the relevant technologies associated with WSN-based SHM systems.

Chapter 2: The Requirements and Design of Wireless Sensor Nodes for SHM. This chapter gives the requirements of SHM applications and an overview of our hardware and software design of wireless sensor nodes.

Chapter 3: Network-Wide and Reliable Event-Triggered Wakeup in WSNs. Wake-up scheduling is an important approach to save energy in WSNs. However, the requirements of wake-up in SHM are different from other applications of WSNs. This chapter provides the details of how to realize network-wide and reliable event-triggered wakeup in WSN-based SHM applications.

Chapter 4: Design of Distributed SHM Algorithms Within Wsns: A Cluster-Based Approach. Another important approach to address the resource-limited WSNs is to embed SHM algorithms within the network. However, typical SHM algorithms are centralized, computationally intensive, and are not easy to be distributed within a network. This chapter provides one typical approach, clustering, to design distributed versions of SHM algorithms.

Chapter 5: Design of Distributed SHM Algorithms Within WSNs: A Networked-Computing Approach. Although the cluster-based approach proposed in Chap. 4 is simple, the accuracy of damage detection obtained may not be guaranteed to be comparable with the centralized one. The goal of this chapter is to design a distributed SHM algorithm which is able to achieve the same accuracy as the centralized counterpart but uses much less wireless transmission cost. We believe the proposed schemes in these two chapters can serve as a guideline for designing distributed SHM algorithms in WSNs.

Chapter 6: Realizing Fault-Tolerant SHM in WSNs. It is well recognized that low cost wireless sensor nodes are likely to exhibit different types of faults, leading to downgrades of system performance. Existing fault-tolerance schemes usually cannot work well in SHM applications because SHM applications are generally data intensive. This chapter provides a series of systematic approaches to address the fault-tolerance issues in a WSN-based SHM system.

Chapter 7: Conclusion and Future Trends. This chapter provides the conclusions of this book.

Acknowledgments

The authors are deeply grateful to the research staff and students in our research group for their hard work in carrying out the ITF project "A Versatile Wireless Sensor Network Platform for Structural Health Monitoring," and the Hong Kong PolyU Niche Area project "Structural Health Monitoring using Wireless Sensor Network." We thank our project group members, including Mr. Yang Liu, Dr. Md Zakirul Alam Bhuiyan, Dr. Chao Yang, and Mrs. Qionglei Hu. We also express our thanks to Prof. Wenzhan Song and Dr. Shaojie Tang for their invaluable advice throughout this research. The financial support from the Innovation and Technology Fund (ITF/39209) of the government of Hong Kong Special Administrative Region, and the NSF of China with Grant 61332004 and 61572218 is greatly appreciated.

Contents

Chapter 1
Introduction

Civil infrastructure systems such as bridges and buildings are expensive assets of our society. Since they are deteriorating with time, to monitor their condition and provide timely alarms is of crucial importance. Traditional systems monitoring structural condition generally leverage cables to deliver data from sensor nodes to an on-site server. These wire-based systems have some intrinsic limitations including high cost and long deployment time. Recently emerged systems using wireless sensor networks (WSNs) start to gain more and more attentions due to their low cost and ease of deployment. However, designing such a wireless system faces some significant challenges that must be carefully addressed. In this chapter, we first introduce the background of structural health monitoring (SHM) and the conventional wire-based SHM systems. Then we describe some existing SHM systems using WSNs. Finally, we summarize the challenges and issues of designing a practical WSN-based SHM system.

1.1 Structural Health Monitoring and Wire-Based Monitoring Systems

Civil structures such as dams and long-span bridges are critical components of the economic and industrial infrastructure. Therefore, it is important to monitor their integrity and detect/pinpoint any possible damage before the damage reaches to a critical state. This is the objective of SHM [2]. In a typical SHM system shown in Fig. 1.1, different kinds of sensors, such as accelerometers and strain gauges, are deployed on the structure under monitoring. These sensor nodes collect the vibration, strain of the structure under different locations, and transmit the data to a central station. Based on the data, SHM algorithms are used to extract damage-associated information and make corresponding decisions about structural condition [2].

© The Author(s) 2016
J. Cao and X. Liu, *Wireless Sensor Networks for Structural Health Monitoring*,
SpringerBriefs in Electrical and Computer Engineering,
DOI 10.1007/978-3-319-29034-8_1

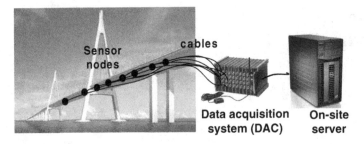

Fig. 1.1 A typical wire-based SHM system

According to the duration of deployment, SHM systems can be largely divided into two categories: short-term and long-term monitoring. Short-term SHM systems are generally used in routine annual inspection or urgent safety evaluation after some unexpected events such as earthquake, overload, or collisions. These short-term systems are usually deployed on structures for a few hours to collect enough amounts of data for off-line diagnosis afterward. Examples of short-term SHM systems can be found in the Humber Bridge of UK [1], and the National Aquatic Centre in Beijing, China [6]. The second category of SHM systems include those used for long-term monitoring. Sensor nodes in these systems are deployed on structures for months, years, or even tens of years to monitor the structures' healthy condition. Different from short-term monitoring systems where data are processed off-line by human operators, most of the long-term SHM systems require the healthy condition of the structure which is reported in a real-time or near-real-time manner. Examples of long-term monitoring SHM systems can be found in the Tsing Ma Bridge and Stonecutters Bridge in Hong Kong [9].

The main drawback of traditional wire-based SHM systems is the high cost. The high cost mainly comes from the centralized data acquisition system (DAC), long cables, sensors, and in-field servers. Particularly for DAC, its price increases dramatically with the number of channels it can accept. As a result, the cost of a typical wire-based SHM system is generally high. For example, the cost of the systems deployed on the Bill Emerson Memorial Bridge and Tsing Ma Bridge reach $1.3 and $8 million, respectively [9].

In addition, deploying a wire-based SHM system generally takes a long period of time. This drawback is particularly apparent in SHM systems used for short-term purpose. Considering the length of cables used in a SHM system deployed on a large civil infrastructure can reach thousands or even tens of thousand meters, deployment can take hours or even days to obtain measurement data just for a few minutes. Moreover, constrained by the number sensor nodes and the capability of DAC, it is quite common that a SHM system is repeatedly deployed in different areas of a structure to implement measurement. This dramatically increases the deployment cost. We have collaborated with civil researches to deploy a wire-based SHM system on the Hedong Bridge in Guangzhou, China (see Fig. 1.2). The DAC system we used can only support inputs from 7 accelerometers simultaneously. To measure the

Fig. 1.2 A wire-based SHM system deployed on the Hedong Bridge, China. **a** Hedong Bridge. **b** Deploying a wired system

vibration at different locations across the whole bridge, the system hence was moved to 15 different areas of the bridge to implement measurement respectively. For each deployment, it took about two hours for sensor installation, cable deployment, and initial debugging.

1.2 Structural Health Monitoring Using Wireless Sensor Networks

Recent years have witnessed a booming advancement of wireless sensor networks (WSNs) and an increasing interest of using WSNs for SHM. Compared with the traditional wire-based SHM systems, wireless communication eradicates the need for wires and therefore represents a significant cost reduction and convenience in deployment. A WSN-based SHM system can achieve finer grain of monitoring, which potentially increases the accuracy and reliability of the system.

We should also notice that SHM is different in many aspects from most of the existing applications of WSNs. Table 1.1 summarizes the main differences between SHM and a typical application of WSNs, environmental monitoring, in terms of sensing and processing algorithms. Sensor nodes in a SHM system typically imple-

Table 1.1 Difference between SHM and environmental monitoring

	SHM	Env. monitoring
Sensor type	Accelerometers, strain gauges	Temperature, light, humidity
Sampling pattern	Synchronous sampling round by round	Not necessarily synchronized
Sampling frequency	X00~X000/sec	X/sec, min
Processing algorithms	On a bunch of data (>X0000) centralized, computationally intensive	Simple, easily to be distributed

ment synchronized sensing with relatively high sampling frequency. Moreover, SHM algorithms to detect damage are based on a bunch of data (in a level of thousands and tens of thousands), and are usually centralized and complicated.

In the remaining of this section, some of the real implementations of WSN-based SHM systems, along with their achievements and limitations, are described.

The Geumdang Bridge

The Geumdang Bridge in South Korea is one of the world's first to employ low-cost wireless sensors. The researchers from University of Michigan and Stanford University collaborated with researchers in South Korea has deployed 14 wireless sensor nodes on the bridge to monitor its response to speeding and overloaded trucks [5]. A laptop computer, with a compatible wireless radio attached, is used to collect bridge response data from the deployed wireless sensor nodes. It should be noted that a hub-spike architecture is adopted in the SHM system which means that the laptop is within the single communication range of all the wireless sensor nodes. The performance of the wireless system is validated by comparing the data quality sampled from a wired system. In addition, fast Fourier transform (FFT) and the peak picking techniques (PP) are implemented in each individual sensor node. The processed information is sent back to the base station and matches well with that obtained from post-processing of data obtained by the wired monitoring system. However, some limitations of the system are still noteworthy. The tested results showed that synchronization error among the 14 sensor nodes can be as large as 0.1 s. Also, single-hop communication architecture may not be appropriate for large structures. Power consumption issue has not been carefully considered and the deployment is only for short-term purpose.

The Golden Gate Bridge

Berkeley researchers installed 64 MicaZ motes with their customized sensor board on the Golden Gate Bridge [4, 7]. The deployed 64 sensor nodes constitute a network with 46 hops. These wireless sensor nodes were sampled at 1 kHz and then averaged and down-sampled at 200 Hz. Sampled data are first stored in the flash on board and then transmitted afterward. To deliver the sampled data reliably through multi-hop wireless communication, they implemented a reliable communication protocol called Straw. In addition, to improve the throughput of the system, a pipelining data transmission technique is adopted which allows nodes within the network to transmit data simultaneously. Despite of the achievement that the authors claimed, one limitation of the system is the bandwidth. Although pipeline technique has been used, it still took over 12 h to complete the transmission of the 20MB of data (1600 s of data, sampling at 50 Hz on 64 sensor nodes) reliably back to a central station. The effective bandwidth achieved is only 3.5 kbps, far below the 250 kpbs theoretical bandwidth of MicaZ. Such a long delivery time not only indicates a serious delay, but also implies significant amount of energy consumption. As a solution, expensive four high-volume lantern batteries (11Ah) are equipped with each sensor node. However, even this battery pack can only support about ten days of continuous work.

The Jindo Bridge

Jindo Bridge is a cable-stayed bridge in South Korea with a 344-m main span and two 70-m side spans. The WSN-based SHM system on the Jindo Bridge is a collaborative work of researchers from South Korea (Korean Advanced Institute of Science and Technology, KAIST), Japan (University of Tokyo), and the USA (University of Illinois at Urbana-Champaign) [3, 8]. They collaborated and deployed a total of 70 Imote2 sensor nodes on the bridge. The primary goal of the Jindo Bridge deployment is to realize the first large-scale, autonomous network of smart sensors utilized for SHM. Particularly, this deployment is to validate the suitability of the Imote2 smart sensor platform, the quality of SHM-A sensor board they have developed, and the performance of the software developed. The WSN-based SHM system deployed on the Jindo Bridge has the similar hub-spoke architecture as the system in the Geumdang but with two base stations. Correspondingly, the network was divided into two subnetworks, one with 37 nodes and the other with 33 nodes. Sensor nodes in each subnetwork are within the single communication range of one of the two base stations. The measured data show a good agreement with data from the existing wired system. An autonomous monitoring is also realized by employing a threshold detection strategy and an energy-efficient sleeping mode (called as 'SnoozeAlarm') to extend the network lifetime. However, the WSN-based SHM system deployed on the Jindo Bridge still has some drawbacks. The single-hop communication network architecture is not appropriate for large structures. Also, when the deployed sensor nodes are in the 'SnoozeAlarm' mode, it takes 1~5 minutes to wake up the entire network. Therefore, the system does not support capturing critical data in short-term, transient events such as an earthquake.

The Torre Aquila Tower

Torre Aquila Tower is a 31-meter-tall medieval tower located in the city of Trento, Italy. Researchers from Italy and Sweden collaboratively designed a WSN-based SHM to monitor the structural response (e.g., deformation and vibration) of the tower, so as to preserve the integrity of the valuable artworks located inside [10]. Since this is a specific application and requires long-term deployment, customized wireless sensor nodes with dedicated communication software have been designed. The WSN-based SHM system installed in the tower consists of 16 sensor nodes, among them are the two fiber optical strain gauges, three accelerometers, and 11 temperature sensors. These sensor nodes, in together with one sink node, are distributed in the five floors of the tower and constitute a multi-hop wireless communication network. The system has been operating since September 2008 for 4 months without changing battery, which is a good performance if compared with other WSN-based SHM systems. Also, the data loss ratio was estimated to be less than 0.01. However, the scalability of the system is validated since only 16 sensors are adopted in the system. In addition, the long lifetime of the system is at the expense of long working interval: only three sensor nodes are equipped with accelerometers and they only work about 6 min every day.

References

1. J.M.W. Brownjohn, M. Bocciolone, A. Curami, M. Falco, A. Zasso, Humber bridge full-scale measurement campaigns 1990–1991. J. Wind Eng. Ind. Aerodyn. **52**, 185–218 (1994)
2. C.R. Farrar, K. Worden, An introduction to structural health monitoring. Philos. Trans. R. Soc. A: Math. Phys. Eng. Sci. **365**(1851), 303 (2007)
3. S. Jang, H. Jo, S. Cho, K. Mechitov, J.A. Rice, S.H. Sim, H.J. Jung, C.B. Yun, B.F. Spencer Jr, G. Agha, Structural health monitoring of a cable-stayed bridge using smart sensor technology: deployment and evaluation. Smart Struct. Syst. **6**(5–6), 439–459 (2010)
4. S. Kim, S. Pakzad, Health monitoring of civil infrastructures using wireless sensor networks, in *Proceedings of the 6th international conference on Information processing in sensor networks* (ACM, 2007), p. 263
5. J.P. Lynch, Y. Wang, K.H. Law, J.H. Yi, C.G. Lee, C.B. Yun, Validation of a large-scale wireless structural monitoring system on the geumdang bridge, in *Proceedings of 9th International Conference on Structural Safety and Reliability* (2005)
6. J. Ou, H. Li, Structural health monitoring in mainland China: review and future trends. Struct. Health Monit. **9**(3), 219 (2010)
7. S.N. Pakzad, G.L. Fenves, S. Kim, D.E. Culler, Design and implementation of scalable wireless sensor network for structural monitoring. J. Infrastruct. Syst. **14**(1), 89–101 (2008)
8. J.A. Rice, K. Mechitov, S.H. Sim, T. Nagayama, S. Jang, R. Kim, B.F. Spencer Jr, G. Agha, Y. Fujino, Flexible smart sensor framework for autonomous structural health monitoring. Smart Struct. Syst. **6**(5–6), 423–438 (2010)
9. K.Y. Wong, Instrumentation and health monitoring of cable-supported bridges. Struct. Control Health Monit. **11**(2), 91–124 (2004)
10. D. Zonta, H. Wu, M. Pozzi, P. Zanon, M. Ceriotti, L. Mottola, G. Pietro Picco, A.L. Murphy, S. Guna, M. Corra, Wireless sensor networks for permanent health monitoring of historic buildings. Smart Struct. Syst. **6**(5–6), 595–618 (2010)

Chapter 2
Requirements, Challenges, and Summary of Hardware and Software Design for a WSN-Based SHM System

Compared to the short-term systems, a typical long-term WSN-based SHM system poses many challenges. In this book, without specifying otherwise, we focus on the design of long-term WSN-based SHM systems. In this chapter, we first describe system requirements. Then we raise a few challenges when designing such a SHM system using WSNs. Finally, we summarize the hardware and software design of our SenetSHM platform.

2.1 Requirements and Challenges

Our collaborators from civil engineering specified the following system requirements for a typical long-term SHM system.

For long-term WSN-based SHM systems, continuous collecting data are neither required nor feasible. It is highly preferable for the system to work only during the occurrence of some certain kinds of events such as earthquake and large wind, etc. In other conditions, sensor nodes are put into sleep to save energy. Thus *event-triggered wakeup* is necessary. Moreover, the wakeup of sensor nodes should be *fast, network-wide, and reliable*. Particularly, the network-wide wakeup means that during the events, all the deployed sensor nodes, even far from the location of event sources, should be awake and start sampling. Second, the system should be able to provide real-time or near-real-time *healthy information* of the structure under monitoring. Last but not the least, a WSN-based SHM system which solely relies on battery should be able to work for weeks or even months.

However, to meet the requirements above entails many challenges. The first one is associated with the network-wide and reliable wakeup in the presence of critical events. In comparison, sensor nodes in [1] work at fixed duty cycle (about 6 min every day), which limits the ability of the application users to initiate network operations at random or can miss the event of interest. The 'SnoozeAlarm' with 'sentry-based' approach in [3] requires constant power supply for sentry nodes. Moreover, the sleep

© The Author(s) 2016
J. Cao and X. Liu, *Wireless Sensor Networks for Structural Health Monitoring*,
SpringerBriefs in Electrical and Computer Engineering,
DOI 10.1007/978-3-319-29034-8_2

sensor nodes are awakened one by one by a central gateway, which leads to a long wakeup time of the network (1–5 min) after the event is first detected. As a result, the system does not support capturing critical data in short-term, transient events such as an earthquake. The system in Brimon [2] is only limited to railway bridge monitoring. And to wake up sensor nodes deployed on the bridge, oncoming trains passing by need to have a node installed to broadcast beacons constantly.

Another issue in long-term SHM is embedding effective SHM algorithms within WSNs. To provide real-time or near-real-time healthy information as well as to save energy, we generally need to implement SHM algorithms within a WSN. However, SHM algorithms used in traditional wire-based SHM systems generally have two properties:

- The SHM algorithms are centralized and require the raw data from deployed sensor nodes.
- The SHM algorithms involve complicated signal processing techniques and require powerful computation units.

The two associated requirements pose significant challenges for a WSN-based SHM system, considering the limited power computational capability (i.e., power CPU and a large memory) of wireless sensor nodes. Some SHM algorithms simply cannot be implemented, or it takes significant amount of time to finish the computation, even longer than transmission the raw data. In addition, we also to need to consider the necessary wireless communication needed for a SHM algorithm.

The last challenge is related to fault tolerance. As was in other WSN applications, wireless sensor nodes deployed in a SHM system can have various types of faults. Moreover, we are interested in detecting faulty sensor readings. Despite of many existing fault-tolerance schemes to address faulty readings in WSNs, they are not able to work well in SHM because it leverages a different model to detect event (i.e., structural damage). Some assumptions in the existing fault-tolerant event detection are not valid in SHM.

2.2 Hardware Design

The hardware architecture of SenetSHM is shown in Fig. 2.1. The SenetSHM platform includes an Imote2 and a specially designed sensor board. Imote2 is chosen as the central unit because it has a good balance between low power consumption and rich resources. Using Imote2, implementing complicated SHM algorithms becomes possible. Imote2 also integrates a radio transceiver CC2420 which will take the main responsibility of transmitting raw data and control commands in SenetSHM.

However, Imote2 alone misses some key components which are necessary for the SenetSHM. Therefore, we design a sensor board that can be attached to the Imote2 to fulfill the following functions:

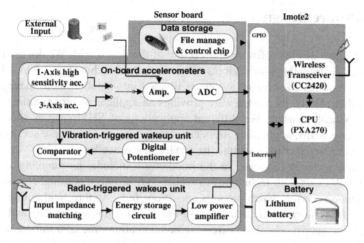

Fig. 2.1 The hardware architecture of the SenetSHM

- **Support for on-board accelerometers and external input**: The sensor board provides abundant support for measuring accelerations. The on-board sensors contain a general-purpose three-axis accelerometer, LIS344ALH, and another high sensitivity one-axis on-board accelerometer SD1221. According to different application scenarios, users can specify which accelerometer will be used through a switch or use them simultaneously. Signals generated from these two sensors are amplified and are transformed to the digital format through a programmable ADC QF4A512. Besides on-board sensors, other types of external sensors such as piezoelectric accelerometers, strain gauges, can also be directly attached with the SenetSHM nodes.
- **On-board data storage**: We also design interfaces on each SenetSHM node for μSD and USB. For each node, the measured data with the associated time stamps are stored into the μSD or the USB in a real-time manner. Both the μSD and the USB used have 2G Byte space, allowing the storage of raw data continuously sampled at 1 KHz for more than two days.
- **Vibration-triggered wakeup and radio-triggered wakeup units**: For long-term SHM, we designed two units, vibration- and a radio-triggered wakeup units and they work together to realize fast and unified wakeup of sensor nodes in the presence of some events. The vibration-triggered wakeup unit will wake up the attached sensor node from deep sleep mode when the vibration of the structure exceeds a pre-defined threshold. On the other hand, the radio-triggered wakeup unit will wake up the sensor node when it receives wakeup messages from others. Different from the 'SnoozeAlarm' mode in [3], SenetSHM nodes with radio-triggered wakeup unit do not need to wake up periodically to listen to the wireless channel and therefore can be more energy efficient and fast. How these two wakeup units collaborate will be described in Chap. 3.

2.3 Software Design

Figure 2.2 illustrates the software architecture of the SenetSHM. We design a middleware for SenetSHM which adopts service-oriented architecture (SOA). Using SOA, the complicated software system is divided into smaller, more manageable services. Particularly, the middleware provides an application programming interface for application users. For different applications of SHM such as short-term and long-term SHM, application users can simply choose from these services and compose them together to constitute the service that is needed.

The basic services contained in the middleware include *the sampling service*, which mainly deals with techniques to realize synchronized sensing; *the wireless communication service*, which supports one-to-one (e.g., for threshold setting), one-to-all (e.g., for time synchronization), all-to-one (e.g., for data collection), and all-to-all (e.g., for wakeup mechanism) wireless communications; *the data storage service*, which allows on-board data storage for short-term SHM; *the wakeup service*, which is used to provide different methods for fast, unified, and synchronized wakeup for long-term SHM applications; *the structural status service*, which provides healthy status of the structure in long-term SHM applications; and the other services such as *the fault-tolerant service* and *other services for maintenance and debug*.

In the following three chapters, we will describe how the SenetSHM addresses the following three important issues: (1) network-wide and event-triggered wakeup, (2) distributed processing of SHM algorithms, and (3) realizing fault-tolerant SHM in WSNs.

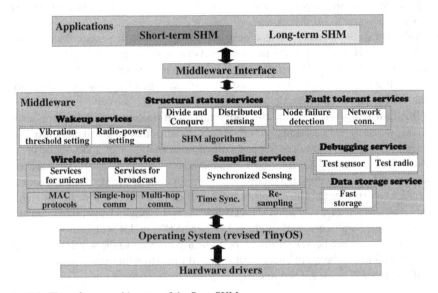

Fig. 2.2 The software architecture of the SenetSHM

References

1. M. Ceriotti, L. Mottola, G.P. Picco, A.L. Murphy, S. Guna, M. Corra, M. Pozzi, D. Zonta, P. Zanon, Monitoring heritage buildings with wireless sensor networks: the torre aquila deployment, in *Proceedings of the 2009 International Conference on Information Processing in Sensor Networks* (IEEE Computer Society, 2009), pp. 277–288
2. K. Chebrolu, B. Raman, N. Mishra, P.K. Valiveti, R. Kumar, Brimon: a sensor network system for railway bridge monitoring, in *Proceedings of the 6th international conference on mobile systems, applications, and services* (ACM, 2008), pp. 2–14
3. J.A. Rice, K. Mechitov, S.H. Sim, T. Nagayama, S. Jang, R. Kim, B.F. Spencer Jr, G. Agha, Y. Fujino, Flexible smart sensor framework for autonomous structural health monitoring. Smart Struct. Syst. **6**(5–6), 423–438 (2010)

Chapter 3
Enabling Network-Wide and Event-Triggered Wakeup

3.1 Introduction

Event-triggered wakeup is widely used in many applications of WSNs such as environmental monitoring, battle field surveillance, forest fire detection, etc. Initially, deployed sensor nodes are put into sleep mode. When some application-specified events such as temperature change or noticeable vibrations occur, nodes near the event location are awaken to collect the information associated with events for application users. Event-triggered wakeup is an important tool to save energy while still fulfilling the tasks required for the system.

To realize event-triggered wakeup in WSNs, many mechanisms have been proposed. For example, each node can be equipped with one or more types of low-power sensors which serve as 'event-detector' and they can wake up the node when necessary [6, 10]. Another example is to put nodes into the 'SnoozeAlarm' mode [4, 12], in which they periodically wake up to listen to the possible messages from some 'sentry nodes' and determine whether to work or continue to sleep accordingly.

In SHM, event-triggered wakeup is particularly useful. Due to the data-intensive nature of this application, it is highly preferable that sensors on a structure collect data only under some certain events such as gusts, vessel hits, earthquakes, etc., since data collected during these events are much more informative for damage detection purpose. However, existing event-triggered wakeup mechanisms mentioned above cannot be directly used, since SHM has different requirements of wakeup from 'conventional' applications of WSNs. To summarize, the wakeup of sensor nodes in SHM applications should be **network-wide, reliable, and fast**.

The requirement of **network-wide** wakeup in SHM is mainly due to the fact that detecting the above events (e.g., a gust of wind, an earthquake) is not the final objective in these applications. In conventional applications of WSNs, the main task for a WSN is to monitor whether some application-specified events occur. Therefore, only nodes which are close to the event locations need to be triggered to work since they contain much more information about the event than others. However, the events

© The Author(s) 2016
J. Cao and X. Liu, *Wireless Sensor Networks for Structural Health Monitoring*,
SpringerBriefs in Electrical and Computer Engineering,
DOI 10.1007/978-3-319-29034-8_3

in SHM only serve as some pre-condition to carry out data collection. The objective of SHM is NOT to detect an earthquake or a gust of wind, but to determine the possible structural damage locations from the data collected during these events. Since damage locations do not necessarily coincide with the event location, data from ALL the deployed sensor nodes on a structure are required, not limited to those near to the event locations. This network-wide wakeup is actually a common requirement in many applications of WSNs but has not yet been fully addressed.

Besides the network-wide wakeup, the wakeup in SHM should be **reliable**. Conventional applications of WSNs such as environmental monitoring generally should avoid false negative 'sleep-in' but allows for some false positive 'wake-ups' since the price for the latter is rather low: each node only needs to collect a few data and can go back to sleep again immediately. However, in SHM, to determine the location of possible damage, each node needs to collect and transmit up to thousands of data, which is rather a time- and energy-consuming task. The problem can become even more serious if network-wide wakeup is required. Therefore, the false wakeups should always be avoided in these applications.

At last, SHM application requires the wakeup should be **fast**. Different from forest fire detection and environmental monitoring where an event can last up to hours, events in these applications such as an earthquake or a gust of wind may last only minutes or seconds.

In this chapter, taking an example of our platform designed for SHM called Senet-SHM, we propose a chain-reaction wakeup mechanism to realize network-wide, reliable, and fast wakeup of wireless sensor nodes in the presence of vibration-associated events. This mechanism leverages two wakeup units and optimally selected sentry nodes to fulfill the task. Compared with the existing event-triggered wakeup mechanism, this method has the following advantages:

- No particular sentry nodes need to be deployed since they are selected from the deployed sensor nodes. In addition, the chain-reaction wakeup mechanism does not require each node is in the direct communication with a sentry node.
- Using this mechanism, 'network-wide, reliable and fast' wakeup can be expected: Once event occurs, the chain-reaction mechanism is able to wake up all the sensor nodes. In addition, the number as well the locations of the sentry nodes are carefully selected to minimize the wakeup delay while still satisfy the pre-defined false positive constraint.

3.2   Related Works

Existing event-triggered wakeup mechanism in WSNs can be divided into two categories: wakeup via low-power sensors and wakeup through wireless communications.

In the first category, each node is equipped with some low-power sensors such as infrared sensors, light sensors, and accelerometers which are able to continuously monitor the events of interest [6]. Once an event is detect, the sensor will wake up the corresponding nodes for further process. Obviously, this approach has small wakeup delay but has difficulty to achieve network-wide and reliable wakeup. For nodes which are far from the event location, the effects of event and environmental noise are hardly distinguishable. It is difficult to determine the thresholds for them with which low false positive 'wake-ups' and low false negative 'sleep-ins' are achieved simultaneously.

In the second approach, sensor nodes utilize wireless communication to wake up each other. Among this approach, a technique called 'SnoozeAlarm' is widely used. In this technique, some specifically designed wireless nodes, generally called as sentry nodes, are deployed. They continuously monitor the events and send wakeup messages when an event is detected. The remaining ones are put into the 'SnoozeAlarm' mode: they periodically wake up to listen for a while whether there are possible messages from the sentry nodes and determine to work or to sleep accordingly. An example of 'SnoozeAlarm' can be found in [12] where 70 Imote2 deployed on the Jindo Bridge are required to collect data in the presence of strong wind. Another example is Brimon [4], where nodes placed in front from trains keep sending beacon signal to trigger the data collection process of sleep nodes on the bridge. This 'SnoozeAlarm' mode is also realized in X-Mac [3] in TinyOS. Some wireless transceivers, such as CC1101 [13], even embed this function in the hardware.

Using this 'SnoozeAlarm' mode for event-triggered wakeup is reliable but also has a few drawbacks. First, special sentry nodes are required which either need continuous power supply or are equipped with special sensors. For example, the sentry nodes used in [12] are equipped with anemometers and require constant power supply. Second, as was pointed in [4], there is always a tradeoff between energy consumption and wakeup delay. In addition, all the deployed sensor nodes need to be within the single-hop communication range with at least one sentry node. This 'single-hop constraint' can have difficulty in a large network.

Besides SnoozeAlarm mode, sensor nodes can also be equipped with specially designed 'radio-triggered circuit' which can collect the energy contained in the wakeup messages sent from the sentry nodes and wake up the corresponding sensor nodes [1, 7]. Different from the 'SnoozeAlarm' mode, a sensor node with radio-triggered circuit does not need to wake up and listen to the wireless channel periodically and therefore can be more energy efficient. However, as in the SnoozeAlarm mode, this approach also requires specially designed sentry nodes and has the 'single-hop' constraint.

There are also some works which incorporate multiple wakeup mechanisms. An example is TelosW [10]. TelosW also includes the CC1101, which has 'SnoozeAlarm' mode, and sensor-initiated wakeup units to wake up nodes by vibrations. TelosW allow users to specify which kind of mechanism is to be used. However, different wakeup mechanisms work independently, and the drawbacks of individual wakeup mechanism are not addressed.

3.3 Preliminaries

In this section, we first briefly introduce the radio- and the vibration-triggered wakeup units designed for event-triggered wakeup in SHM. Then we define some concepts required for the problem formulation in the next section. Table 3.1 summarizes the notations used in this chapter.

3.3.1 Two Wakeup Units and the Chain-Reaction Wakeup

Similar to the sensor-initiated wakeup used in [6, 10], we designed a vibration-triggered wakeup unit with which a SenetSHM node is able to be awake when the measured vibration exceeds a pre-defined threshold. Thus the wakeup becomes automatic and fast. The unit contains a comparator, which compares two inputs, one set by the node as a reference, and the other from the analog accelerometer. If the value from the accelerometer exceeds the reference one, the comparator generates an interrupt to wake up the node.

As a supplement to the vibration-triggered unit, we design a radio-triggered wakeup unit similar like the 'radio-triggered circuit' used in [1, 7]. A radio-triggered wakeup unit is able to generate interrupt to wake up the corresponding node once it collects enough energy contained in wireless packets sent from others. The power consumption of both wakeup units are also less than 1 mA.

In this chapter, all the deployed nodes are equipped with radio-triggered wakeup units, while the vibration-triggered wakeup units are only enabled on part of the nodes called as sentry nodes. Initially, all the nodes are put into the sleep mode. When an event occurs, the sentry nodes will be woken up by their vibration-triggered units

Table 3.1 Summary of notations

p_i^-, p_i^+	Node-wide false negative and positive ratio of node v_i
τ_i	The threshold set on the vibration-triggered wakeup unit of node v_i
$\mathcal{P}^+(S), \alpha$	The network-wide false positive ratio and its threshold
δt_i	The period of signal from node i when event occurs
T_i^{ev}	The expected time for the vibration-triggered wakeup unit of v_i to be woken up by a wavelet of event
T^{ini}	The time it takes for a node to be initialized
T_{ij}	The time between v_i broadcasting its 1st msg to an interrupt generated at v_j
t_{ij}	The time starting from v_i sending its 1st msg to v_j, to v_j is ready to send its 1st msg to its neighbors
t_i	The average time it takes for an event to wake up v_i
V_{ori}, V_{vir}	Nodes in the sensor layer and the event layer, resp.
M	The number of maximal feasible sets generated in the R–K

first. They then broadcast the wakeup message to their neighbors. A node which is woken up will continue to broadcast the wakeup message. This chain reaction will continue until all the nodes are awake.

Obviously, the number as well as the locations of sentry nodes will cause different wakeup delays and reliabilities. For example, a WSN with a large number of sentry nodes is likely to have small wakeup delay but many false wakeups. How to select the sentry nodes such that the wakeup delay is minimized under a pre-defined constraint on false wakeups is the problem we aim to solve. Before we formulate this problem, we first give some preliminaries.

3.3.2 Event, False Positive, and False Negative

In this section, we first give the definition of the event and the node-wide false positive ratio p_i^+ and false negative ratio p_i^-. Then we describe how they affect the network-wide false positive ratio and the wakeup delay, respectively.

Event Definition

Typical events of interest to SHM include a gust of wind, an earthquake, a vessel hit, etc. During these events, data measured at nodes deployed on the structure can be represented as periodic signals. As an example, Fig. 3.1 shows the data collected during 'a gust of wind' event from our SenetSHM nodes deployed on the Hedong Bridge in Guangzhou, China. Figure 3.1a, b shows the data from the sensor nodes on the bridge railing and a cable, respectively. For a certain node, the vibration period of its measured signal is determined by a intrinsic property of the structure called **natural frequencies** as well as the node location. Each structure has a number of

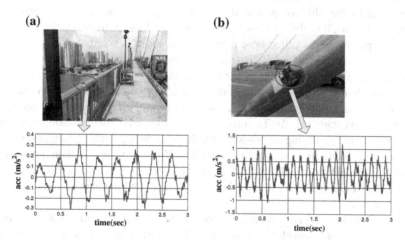

Fig. 3.1 During a gust of wind event, data from **a** a sensor on the bridge railing, and **b** a sensor on a cable

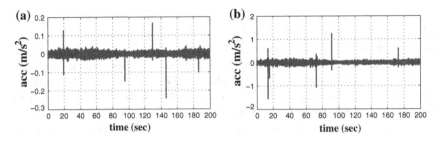

Fig. 3.2 Data collect when no event occurs from **a** the sensor on the railing, and **b** the sensor on the cable

natural frequencies and the measured signal from a node generally is dominated by one of them. It can be seen from Fig. 3.1 that in this event, measured signal at bridge cable has higher vibration frequency and also higher peaks than the one on the deck.

When no defined events occur, the vibration of the structure usually has much less amplitudes caused by environmental noise, traffic load, etc. As illustration, Fig. 3.2 shows the data measured from the two sensors when no wind event occurs. Obviously, the structure is not still, and the peaks in the signals are caused by vehicles on the bridge.

Node-Wide False Positive/Negative

Let H_0 and H_1 denote the hypothesis of the nonoccurrence and the occurrence of the event, respectively. Each node has a threshold τ_i for its vibration-triggered wakeup unit such that when the measurement data exceed the threshold τ_i, it will be awake. When H_1 is true, the signal measured at each node can be regarded as being composed of a series of equally sized wavelets. Each wavelet takes two adjacent troughs in the signal as its start and end points, respectively, and the length of the wavelet is determined by the vibration period of the signal. Due to the variation of amplitudes of these wavelets, it is always possible that a wavelet fails to trigger up the vibration unit. This probability is called false negative ratio and is denoted as p_i^-.

On the other hand, when H_0 is true, the measured signal can also erroneously wake up the unit. Obviously, the larger the time duration of the measured signal, the more possible that an erroneous wakeup occurs. To evaluate it in a more objective way, we define a fixed time slot Δt, and determine the probability that an erroneous wakeup occurs in a Δt. The probability is called false positive ratio p_i^+. In this chapter, without specifying otherwise, Δt is 20 s. According to this setting, the signal shown in Fig. 3.2a can be divided into 10 slots. If we let $\tau_i = 0.1$, then among the 10 slots, 2 contains the peaks exceeding the threshold, and then the $p_i^+ = 0.2$. Practically, the vibration signal sampled during H_0 should be long enough to accurately determine p_i^+.

It should be noted that p_i^- is evaluated on a wavelet but p_i^+ is on a pre-defined time slot. The length of a wavelet is generally different from the time slot. In addition, the p_i^+ and p_i^- are determined not only by the amplitude of measured signals, but also by the threshold τ_i. For a certain sensor node v_i and its measured signals in H_0

and H_1, the higher the τ_i, the smaller the p_i^+, but the larger the p_i^-. In this chapter, when calculating p_i^+ and p_i^-, we assume that the threshold is already set for each node in some way. How to select appropriate τ_i will be described using an example in Sect. 3.6.2.

Network-Wide False Positive Rate and the Vibration Wakeup Delay
p_i^+ only represents node-wide false positive ratio. However, what we are more interested is from the entire system point of view. In this chapter, as long as one sentry node wakes up, the whole network will be awakened by the wakeup chain reaction. Therefore, given a set of sentry nodes $S = \{s_1, \ldots, s_m\}$, the network-wide false positive ratio, denoted as $\mathscr{P}^+(S)$, can be calculated as

$$\mathscr{P}^+(S) = 1 - \prod_{i=1}^{m}(1 - p_i^+) \tag{3.1}$$

In this chapter, we require that $\mathscr{P}^+(S) \leq \alpha$ where α is a pre-defined threshold.

On the other hand, p_i^-, the node-wide false negative rate of node v_i directly affects the time it takes to detect the event. Let δt_i be the length of each wavelet in the signal measured at v_i when H_1 is true, and the expected time it takes for the vibration-triggered unit of v_i to be woken up by the event, denoted as T_i^{ev}, can be calculated as

$$T_i^{ev} = \delta t_i p_i^-(1 - p_i^-) + 2\delta t_i (p_i^-)^2(1 - p_i^-) + \cdots$$
$$= \sum_{k=1}^{\infty} k * \delta t_i * (1 - p_i^-)(p_i^-)^k = \frac{\delta t_i p_i^-}{1 - p_i^-} \tag{3.2}$$

It can be seen from Eq. 3.2 that in the extreme case when $p_i^- = 0$, $T_i^{ev} = 0$, while when $p_i^- = 1$, $T_i^{ev} = \infty$.

3.3.3 Wakeup Graph and Wakeup Delay

In this section, we describe how to calculate the **system wakeup delay** in the chain-reaction mechanism, defined as the time beginning from the event starts till the last node in the network is awake. We start with a simple example as shown in Fig. 3.3 to illustrate the breakdown of the time spent in the wakeup mechanism where the system only contains two nodes, one sentry node v_i and the other common node v_j.

Initially, both nodes are in the sleep mode. The time it takes for the vibration wakeup unit of v_i to be woken up by the vibration of the structure is T_i^{ev}, and is calculated as Eq. 3.2. The vibration wakeup unit then generates an interrupt which triggers the initialization of power system of v_i, followed by the booting up of the TinyOS, and the initialization of the sensor board. The total time cost in these three stages, denoted as T^{ini}, is a platform-dependent value. In our SenetSHM platform,

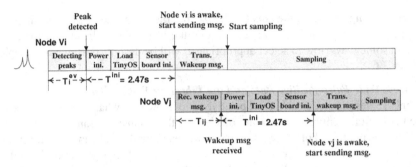

Fig. 3.3 The breakdown of the time spent in the wakeup mechanism

the first two steps take about 2.45 s, while the last step takes about 20 ms, making $T^{ini} \approx 2.47$ s.

As shown in Fig. 3.3, after v_i is awake, it immediately broadcasts a number of wakeup messages to v_j and then starts sampling. From node v_j perspective, an interrupt can be generated once its radio wake up unit accumulates enough energy from messages sent from node v_i. Therefore, T_{ij}, the time starting from v_i broadcasting its first message and ending with an interrupt being successfully generated at v_j, is dependent on the distance and the wireless environment between the two nodes. v_i will then be initialized, and initialization time is also T^{ini}. Therefore, the total wakeup delay of this system, denoted as Del, can be calculated as

$$Del = T_i^{ev} + 2 * T^{ini} + T_{ij} = \frac{\delta t_i p_i^-}{1 - p_i^-} + 2 * T^{ini} + T_{ij} \qquad (3.3)$$

Note that Del is an expected value because of the T_i^{ev}.

Calculating Del is much more difficult when multiple sentry nodes are selected from a WSN. To determine the wakeup delay for a general WSN, we introduce the concept of **wakeup graph**. The wakeup graph, denoted as $G_w = (V, E)$, is a directed graph. V includes two types of nodes: the sensor nodes deployed on the structure and the same number of virtual nodes. Each sensor node has a virtual node and they are located at two different layers, namely, the **sensor layer** and the **event layer**, respectively. An example of wakeup graph is shown in Fig. 3.4, which includes eight sensor nodes and their virtual nodes. For convenience, the sets of nodes in the sensor layer and event layer are denoted as V_{ori} and V_{vir}, respectively. E also includes two types of edges: edges between nodes in V_{ori}, and those connecting nodes from V_{vir} to V_{ori}. For any $v_i, v_j \in V_{ori}$, there exists an edge from v_i to v_j if v_i can wake up v_j using wakeup messages. The weight of the edge from v_i to v_j, denoted as t_{ij}, is the time starting from v_i sending its first message to v_j, to v_j is ready to send its first message to its neighbors. It can be seen from Fig. 3.3 that

$$t_{ij} = T_{ij} + T^{ini} \qquad (3.4)$$

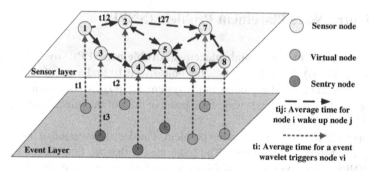

Fig. 3.4 A wakeup graph example

For any node $v_i \in V_{ori}$, there is a corresponding virtual node $v_i' \in V_{vir}$ and an edge from v_i' to v_i. The weight of this edge is denoted as t_i and is the average time it takes for an event to wake up node v_i. According to Fig. 3.3, it can be seen that

$$t_i = T_i^{ev} + T^{ini} = \frac{\delta t_i p_i^-}{1 - p_i^-} + T^{ini} \tag{3.5}$$

Given a wakeup graph $G_w = (V, E)$ and a set of vertices $S \subseteq V_{vir}$ which are the selected sentry nodes to initiate the wakeup process. The **wakeup delay**, denoted as $Del(G_w, S)$, is determined by the maximum distance between a node in V_{ori} with its nearest node in S. That is,

$$Del(G_w, S) = \max_{v \in V_{ori}} \min_{s \in S} d(v, s), \tag{3.6}$$

where the $d(v, s)$ is the accumulated weights along the shortest path from node s to v. Note using Eq. 3.6 to determine the wakeup delay does not consider the possible back-off delay in MAC when multiple nodes try to send messages simultaneously. This is, however, a valid assumption in our system since the radio-triggered wakeup unit does not require wireless message is successfully received. In addition, the MAC protocol in the chain-reaction wakeup mechanism is modified and allows a node to directly send wakeup messages without using the CCA (Clear Channel Assessment).

In practice, given $G_w = (V, E)$ and S, $Del(G_w, S)$ is calculated by first establishing the shortest path trees (SPTs) rooted at each node in S, respectively. In each SPT, every node $v \in V_{ori}$ is labeled with its distance to the root. Then for all the nodes $v \in V_{ori}$, the maximum value among all the SPTs is used as the final one.

So far, the wakeup graph is only used to calculate the network-wide wakeup delay when H_1 is true. To incorporate the constraint of $\mathscr{P}^+$, we require that each virtual node $v' \in V_{vir}$ has the same false positive ratio with its corresponding sensor node in V_{ori}. A wakeup graph hence contains all information required to simulate the wakeup process in the chain-reaction wakeup mechanism.

3.4 Sentry Node Placement Problem (SNPP)

Now we formulate **the sentry node placement problem (SNPP)**: Given the wakeup graph $G_w = (V, E)$, $V = V_{ori} \bigcup V_{vir}$, the objective is to find a set of sentry nodes $S \subseteq V_{vir}$, such that the wakeup delay of the system, $Del(G_w, S)$, is minimized, under the constraint that the network-wide false positive ratio of the selected sentry nodes is $\mathscr{P}^+(S) \leq \alpha$.

The above SNPP is an optimization problem. The decision version of the problem can be proved to be NP-complete. The basic idea is that we first choose a special case of SNPP: when the node-wide false positive ratio of each node is a fixed value σ (i.e., $p_1^+ = p_2^+ = \cdots = \sigma$), and prove that even this special case is NP-complete by reducing the k-center problem to it. Due to the page limit, the detailed proof can be found in [9].

3.5 Solution

We first design two simple greedy heuristic algorithms to solve the SNPP. The first algorithm, called as '**Greedy #1**,' sorts the nodes in V_{vir} according to their p_i^+ and from small to large, adding nodes one by one into the selected sentry set S until the $\mathscr{P}^+(S) \leq \alpha$ is not satisfied. This algorithm can maximize the number of sentry nodes selected and generally works well if nodes with small p_i^+ are spread uniformly over the whole network. In the second algorithm called as '**Greedy #2**,' each time, we add a node $v \in V_{vir}$ into S which is able to maximize the decrease in the delay (i.e., $\max_{v \in V_{vir}}(Del(G_w, S) - Del(G_w, S \cup \{v\})))$ while still satisfying that $\mathscr{P}^+(S \cup \{v\}) \leq \alpha$. The iteration stops until no more v can be added into the S. This algorithm can perform better if the variation of p_i^+ among different nodes is small.

In the remaining of this section, we introduced a novel algorithm, called as Randomized Knapsack (**R-K**). The R-K utilizes the solutions to the Knapsack problem to randomly generate a number of feasible solutions to the SNPP. Through simulation, we will show that this R–K algorithm performs much better than the two greedy algorithms.

The form of the SNPP is similar to the Knapsack problem for which there already exist many solutions such as in [5, 8]. However, directly using these approaches for this problem is not feasible because neither the objective function $Del(G_w, S)$ nor the constraint $\mathscr{P}^+(S)$ are linear. Before we solve the problem, we proposed a few definitions.

Definition 1 (*Feasible sets*) Given a wakeup graph $G_w = (V, E)$, where $V = V_{ori} \bigcup V_{vir}$, a feasible set for the SNPP is a set $S \subseteq V_{vir}$ such that $\mathscr{P}^+(S) \leq \alpha$ is satisfied.

Definition 2 (*Maximally Feasible sets (MaxFS)*) Given a wakeup graph $G_w = (V, E)$, where $V = V_{ori} \bigcup V_{vir}$, a *maximally* feasible set (MaxFS) is any feasible set $S \subseteq V_{vir}$ such that $S \cup \{v\}$ is no longer feasible for any $v \in V_{vir} \backslash S$.

Lemma 3.1 *If $S \subset S'$ and both S, S' are feasible, S' is always the better selection.*

Proof According to Eq. 3.6, it can be clearly seen that adding a sentry node into S will not increase the wakeup delay: $\max\limits_{v \in V_{ori}} \min\limits_{s \in S \cup \{x\}} d(v, s) \leq \max\limits_{v \in V_{ori}} \min\limits_{s \in S} d(v, s)$ for any $x \in V_{vir}$. Therefore, for feasible sets S and S' and $S \subset S'$, $Del(G_w, S') \leq Del(G_w, S)$.

Therefore, we have the following lemma:

Lemma 3.2 *An optimal solution to the SNPP is always a MaxFS.*

In this chapter, we generate the MaxFSs by solving a Knapsack problem. Consider the Knapsack problem: To n objects $X = \{x_1, x_2, \ldots, x_n\}$ we assign profits $r(x_1), \ldots, r(x_n)$, and weights $w(x_1), \ldots, w(x_n)$. With weight constraints w_0, the goal is to find a subset Y of X that maximizes the profit $\sum_{x \in Y} r(x)$, subject to $\sum_{x \in Y} w(x) \leq w_0$.

To generate the MaxFSs of SNPP, the constraint $\mathscr{P}^+(S) \leq \alpha$ of the SNPP is first converted to nonnegative linear constraints by taking logarithms and multiplying by -1:

$$- \log(1 - \mathscr{P}^+(S)) = -\sum_{i \in S} \log(1 - p_i^+) \leq -\log(1 - \alpha) \tag{3.7}$$

The above transformed constraint of SNPP resembles the constraint of the Knapsack problem (i.e., $\sum_{x \in Y} w(x) \leq w_0$). We then generate a two-point random variable $r(x) \in \{\varepsilon, 1\}$ for all $x \in V_{vir}$ where $0 < \varepsilon < 1$ is a pre-defined positive small value. Taking $\sum r(x)$ as the objective function, we have a new problem which exactly matches the Knapsack problem:

Given: n objects $V_{vir} = \{x_1, x_2, \ldots, x_n\}$, with values $r(x_1), \ldots, r(x_n)$ where $r(x_i) \in \{\varepsilon, 1\}$, weights $w(x_1), \ldots, w(x_n)$ where $w(x_i) = -\log(1 - p_i^+)$, and weight constraint $w_0 = -\log(1 - \alpha)$,

The goal: is to find a subset S of V_{vir} that maximizes the profit $\sum_{x \in S} r(x)$, subject to $\sum_{x \in S} w(x) \leq w_0$.

Obviously, this new problem can be solved as a Knapsack problem and the solution is a feasible set for the SNPP. Furthermore, we have following lemma:

Lemma 3.3 *The solution of the Knapsack problem is always a MaxFS of the SNPP.*

Proof Assume the optimal solution to the above problem, denoted as S, is not a MaxFS, then according to the definition of the MaxFS, there exists a $x_k \in V_{vir}$ such that $S \cup \{x\}$ is a feasible set. Considering the fact that $r(x_k) + \sum_{x \in S} r(x) > \sum_{x \in S} r(x)$, then $S \cup \{x\}$ is a better solution than S, which contradicts the assumption.

The purpose of using random $r(x)$ is to generate *different* MaxFSs. Intuitively, if we set $r(x)$ differently each time and repeat the simulation, different MaxFSs can be obtained since in one realization, and nodes in the set $V_1 = \{x | r(x) = 1\}$ are more likely to be selected than nodes in $V_0 = \{x | r(x) = \varepsilon\}$. It can be easily proved that if the SNPP only contains two MaxFSs, namely #1 and #2, and if

$$|MaxFS_1| - |MaxFS_1 \cap V_1| < |MaxFS_2| - |MaxFS_2 \cap V_1|, \qquad (3.8)$$

the MaxFS #1 will be generated in this simulation, where $|MaxFS_1|$ and $|MaxFS_2|$ are the size of the MaxFS #1 and #2, respectively. Similarly, the MaxFS #2 will be generated if otherwise. This conclusion can be generalized to the condition where exist multiple overlapped MaxFSs.

In our method, we generate, for each $x \in V_{vir}$, a $r(x)$ which is a two-point random variable Z_x where $P(Z_x = 1) = c_x$, and $P(Z_x = \varepsilon) = 1 - c_x$, and then solve the Knapsack problem as above. The choice of c_x can be seen as applying a heuristic, as it will favor the inclusion of certain elements in the solution; here, we only consider the case when $c_x = 0.5$ for all x. After we generate M maximally feasible sets, the best one is chosen according to $Del(G_w, S)$. This R–K method is shown in Algorithm 1.

Algorithm 1 Randomized Knapsack (R–K) for the SNPP

$MinDel \leftarrow \infty, S^* \leftarrow \phi$
for $i = 1 \to M$ **do**
  for $x \in V_{vir}$ **do**
    $r(x) \leftarrow$ Two-point random distribution with $c_x = 0.5$
  end for
  Solve the Knapsack problem with:
  - Profits $r(x)$
  - Weights $w(x) = -\log(1 - p_x^+)$
  - Maximum total weight $w_0 = -log(1 - \alpha)$
  $S \leftarrow$ Solution to the Knapsack
  if $Del(G_w, S) < minDel$ **then**
    $MinDel = Del(G_w, S), S^* = S$
  end if
end for
if $S^* = \phi$ **then**
  Return 'Infeasible'
else
  Return S^*
end if

3.6 Simulation

In this section, we describe the simulation results of the three methods proposed in the previous section. The first simulation is based on randomized spanning trees. The purpose of this simulation is to test their performance in different wakeup graphs. In the second simulation, these methods are tested using data from a simulated suspension bridge.

3.6.1 Spanning Tree-Based Simulation

Testing Scenario

The reason we choose spanning trees is because they have the worst optimal wakeup times, as simply adding an edge to a wakeup graph will decrease the expected wakeup time given any selection of sentry nodes. The testing scenario is as follows:

1. We randomly generate a spanning tree $G = (V, E)$ of fixed size (simply by randomly connecting pairs of adding edges between vertices until a spanning tree is obtained). The weight of each edge in E is determined using $t_{ij} = T_{ij} + T^{ini}$, where T_{ij} is set to be a uniformly distributed random variable on the interval $[1/90-1/5]$. This approximately corresponds to the packet loss ratio ranging $10-95\%$ when the practically maximum transmission rate in Zigbee protocol, 100 packets/sec, is used for transmitting wakeup messages. Furthermore, we assume that $t_{ij} = t_{ji}$.

2. For each node in V, we generate a virtual node and assign a p_i^+ and a p_i^- to it. Both p_i^+ and p_i^- are taken uniformly at random from $[0-0.2]$. The weight of edge $t_i = \frac{\delta t_i p_i^-}{1-p_i^-} + T^{ini}$ (see Eq. 3.5). To reflect the practice, δt_i is set to be a uniformly distributed random variable on the interval $[0.1-1]$.

3. After the two procedures described above, we have obtained a wakeup graph. To reflect practical conditions, we let $\alpha = 0.2$. We repeat the experiment for 100 times, and compare the results of different algorithms from statistical point of view.

Simulation Results

First, we compare the performance of three proposed algorithms with the bruteforce search in which the optimal solution is obtained by enumerating all feasible solutions. Due to the nature of this testing approach, it is only feasible to test the algorithms against the optimal solution for small networks. We let the number of nodes $m = 15$. Let T^{grel}, T^{gre2}, T^{R-K}, and T^{bru} be the wakeup delay calculated using the Greedy #1, the Greedy #2, the R-K method, and the bruteforce, respectively. Figure 3.5a–c

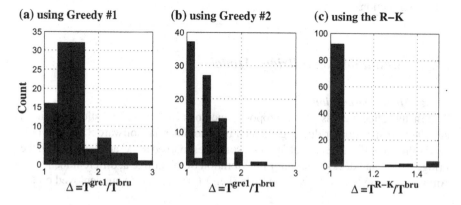

Fig. 3.5 Histogram of simulation results of the three algorithms

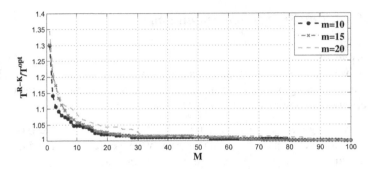

Fig. 3.6 The results from R–K with M under different m

shows the histograms of simulation results of three algorithms, respectively. The x-axes in the three subfigures show the ratio of delays of the three algorithms to the optimal one from the bruteforce, denoted as T^{gre1}/T^{bru}, T^{gre2}/T^{bru}, and T^{R-K}/T^{bru}, respectively. The y-axis is the count. It can be seen that among the three methods, the R–K method is obviously the best: Among 100 tests, it achieves 92 optimal, and the worst-case solution is less than 1.4 times of the optimal one. For Greedy #1 and #2, the former is better since its histogram skews more to the left than the latter.

For the R–K method, its results depend on M, the number of MaxFSs generated. Obviously, the larger the M, the better the result will be. We further test the performance of the R–K algorithm at different M under different network sizes. Initially, the network size $m = 10$, and we let M increase from $1-100$. For each M, we test the R–K method on 100 random graphs and calculate the average of T^{R-K}/T^{bru} for these 100 graphs. Note that T^{bru} is different for different graphs. Then the above procedures are repeated for different m. Figure 3.6 shows how the average of T^{R-K}/T^{bru} changes with M under different m. It can be seen that when $m = 10$, the T^{R-K}/T^{bru} drops sharply with the increase of M and then stabilized soon. In these simulations, M only needs to be a small value for the R–K method to achieve a fairly good result (e.g., $T^{R-K}/T^{bru} < 1.1$ if $M > 10$). In addition, with the increase of m, the M that can achieve a good T^{RK}/T^{opt} does not show a noticeable increase, which might indicate a good scalability of the R–K algorithm. Hereinafter, the $M = 30$ are adopted in the following sections.

3.6.2 Simulation on a Bridge Model

Bridge Model Description

To test the effectiveness of the proposed methods in a more realistic scenario, a simulated suspension bridge is generated by a commercial software SAP2000 [15] and is shown in Fig. 3.7. A total of 36 sensor nodes marked as 'x' are deployed on the bridge cables and decks to monitor the vibration of the bridge at different locations. Among these 36 nodes, 18 with odd IDs (i.e., #1, #3, ..., #35) are deployed on the

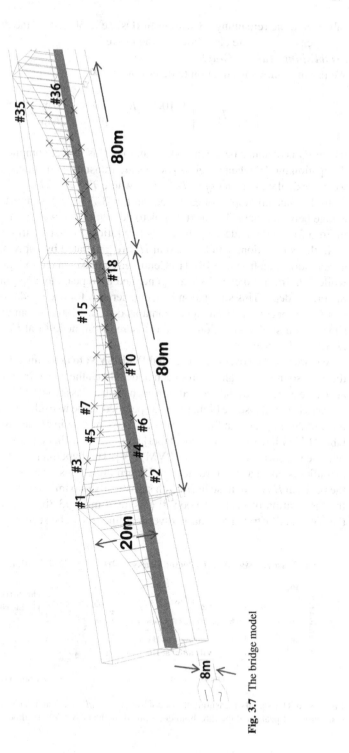

Fig. 3.7 The bridge model

cables, and the remaining 18 with even IDs are deployed on the decks. The IDs of some sensor nodes are also shown in the figure.

Establishing Wakeup Graph

We use the following equation to determine the T_{ij}:

$$T_{ij} = \begin{cases} 1/1000 * R_{ij}^2 & \text{if } R_{ij} \leq 20\,\text{m} \\ \inf & \text{if } R_{ij} > 20\,\text{m} \end{cases} \tag{3.9}$$

where R_{ij} is distance between node v_i and v_j. This model complies with the energy dissipation model when wireless packets are transmitted in an open space [2]. Then we can calculate t_{ij} using $t_{ij} = T_{ij} + T^{ini}$, where $T^{ini} = 2.47\,\text{s}$.

In the wakeup graph, for each node in Fig. 3.7, we create a virtual node and add a edge between them. The next step is to determine the weight of each edge (i.e., t_i in Eq. 3.5) and the node-wide false positive ratio p_i^+ of each virtual node.

In this simulation, data from event H_1 are generated by applying the earthquake acceleration data from the 1940 El Centro earthquake on the four piers of the bridge, while data from 'no event' H_0 are generated by repeatedly applying traffic load on the bridge deck. This simulation hence generates, for each node, two signals in H_0 and H_1, respectively. Each signal contains data in 1200 s. As an example, Fig. 3.8a shows the first 60-s vibration signal measured from node #1 at H_0 (solid curve) and H_1 (dashed curve).

We proposed a novel method called **Peak-SVM** to determine the threshold τ_i and the corresponding p_i^- and p_i^+ for each node. Given the signal from node v_i measured at H_0, we extract the peaks and assign each peak a class label '0.' Note that each peak extracted should also be higher than a specified value to exclude low-amplitude and noise-induced peaks. In the similar way, peaks of data in H_1 are assigned with class label '1.' With these {peak-value, label} pairs, we take them as training data and feed into the support vector machine (SVM) [14]. The hyperplane in the obtained SVM classifier is taken as the threshold τ_i. Determining τ_i using SVM is able to separate the peaks in H_0 from those in H_1 with statistically maximum margin. As an example, the distribution of peaks at node #1 is shown in Fig. 3.8b. The threshold τ_1 using the SVM is illustrated as a dashed vertical line. It can be seen that the distributions

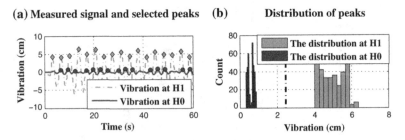

Fig. 3.8 a The signals measured at node #1 at H_0 (*solid curve*) and H_1 (*dashed curve*). **b** The distribution of peaks and the threshold determined by the peak-SVM method

of peaks of node #1 at H_0 and H_1 are distributed in two different regions and the threshold τ_i is able to separate them with the maximum margin.

Based on the τ_i, we can then calculate the false negative p_i^- and false positive p_i^+. Here, we estimate p_i^- using $p_i^- = n_b^{H_1}/n^{H_1}$, in which n^{H_1} is the total number of peaks extracted from signal in H_1 and $n_b^{H_1}$ is those which fall *below* the τ_i. On the other hand, $p_i^+ = n_a^{H_0}/n_{slot}$, where $n_a^{H_0}$ is the number of peaks extracted from signal in H_0 which are *above* the τ_i, and n_{slot} is the number of time slot contained in the signal. It can be seen from Fig. 3.8b that in for node #1, both its n^{H_1} and $n_a^{H_0}$ are zeros. Therefore, $p_i^+ = p_i^- = 0$ for node #1.

In contrast, the first 60-s data measured from node #2 at H_0 and H_1 are illustrated in Fig. 3.9a. The distributions of the peaks extracted from the whole 1200-s data, along with the threshold using the SVM, are shown in Fig. 3.9b. It can be seen that there is an overlap between the peaks in H_0 and H_1. Figure 3.9c shows the whole 1200-s data in H_0 and H_1, and the false positive peaks and false negative peaks classified by the threshold shown in Fig. 3.9b. In particular, among the 459 peaks in H_1, 75 are fall below the threshold, making the p_i^- be approximately 0.16. For p_i^+, since there are only 6 peaks which are above the threshold, and 1200-s data contain a total of 60 time slots each with 20-s length, $p_i^+ = 0.1$ for node #2.

Figure 3.10 illustrates the p_i^+ and p_i^- of nodes on the cables and deck, respectively. The summation of p_i^+ and p_i^- can largely reflect how the peaks of node v_i in H_0 and H_1 can be separated. It can be seen that the peaks extracted from nodes on the bridge cables generally have much different amplitudes in H_0 and H_1 than nodes on the deck.

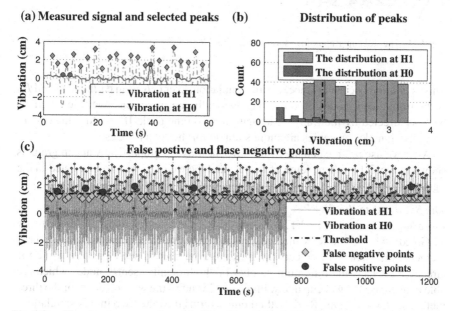

Fig. 3.9 a The signals measured at node #2 at H_0 and H_1. **b** The distribution of peaks and the threshold determined by the peak-SVM method. **c** The false positive and negative peaks

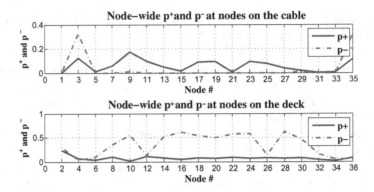

Fig. 3.10 The p_i^+ and p_i^- of nodes on the cables (*upper*) and deck (*lower*)

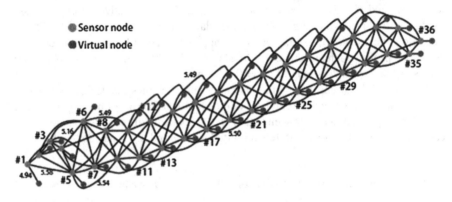

Fig. 3.11 The wakeup graph of sensor nodes in Fig. 3.7

This is intuitively correct since, in this simulation, event H_1 is an earthquake and H_0 is the traffic load, nodes on the cables generally are more sensitive to earthquake but less affected by the traffic load than those on the deck. If no other factors are considered, nodes on cables are more suitable for the sentry nodes.

After we use p_i^- to calculate t_i, we have the wakeup graph. The wakeup graph is shown in Fig. 3.11, where some of the IDs and the weights are also illustrated in the figure.

The Solutions to the SNPP

From top to bottom, Fig. 3.12a shows the selected sentry nodes in the Greedy #1,Greedy #2, and the R–K methods, respectively. It can be seen that among these methods, the Greedy #1 selects the maximum number of sentry nodes, while the Greedy #2 selects the minimum number of 'high quality' sentry nodes which are able to decrease the wakeup delay. Figure 3.12b shows the wakeup delay of the three methods. Obviously, the R–K method outperforms the other two in this simulation.

(a)

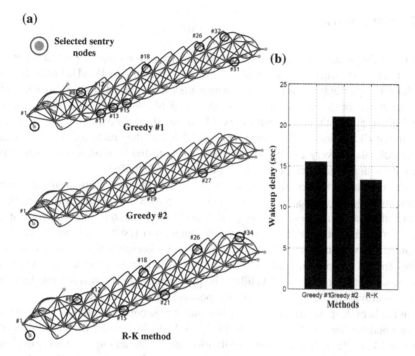

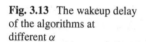

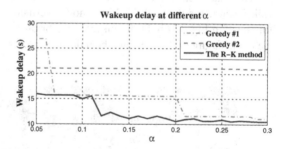

Fig. 3.12 **a** The sentry nodes selected in the three methods. **b** The corresponding wakeup delay

Fig. 3.13 The wakeup delay
of the algorithms at
different α

Figure 3.13 shows the wakeup delay of the three methods when a different α (i.e., false positive upper-bound) is adopted. Intuitively, the minimum wakeup delay is a nonincreasing function of α since a larger α potentially allows more sentry nodes to be added. We can see that when α is increasing from 0.05 to 0.3, except Greedy #2, the wakeup delay of the other two methods drops significantly. Again, the R–K method performs the best in all possible α. In addition, we can see that compared with the Greedy #1, the wakeup delay of the R–K method drops more gradually, which reflects its superior capability to approaching to the optimal value. The slight increase of wakeup delay at some points is because of the sub-optimal results of the R–K method, which shall disappear with larger M.

3.7   Experiment

We use a simple experiment to test the performance of the proposed wakeup mechanism. The test structure is the LSK building in Hong Kong PolyU. From floor 9 to floor 13, we deployed a total of 9 sensor nodes, numbered from #1 to #9, each on one corner of the stairs. As an example, Fig. 3.14 shows the nodes deployed on the 10th, 10.5th, 11th, and 11.5th floors. The event H_1 is generated by two persons, one in the 10.5th floor and the other in the 12.5th, and they raise and drop one heel repeatedly about every 2 s. While 'no event' H_0 is tested by a person continuously walking with normal pace between 9th floor and the 13th floor.

We summarize the testing procedures as follows: we first carried out two tests including about 2-min heel drop (i.e., H_1) followed by 8-min normal walk (i.e., H_0). Data collected during these tests are used to estimate p_i^+ and t_i for the wakeup graph. We then carried out another wireless communication test to find out t_{ij}. Having obtained the wakeup graph, we use the proposed algorithms to choose the sentry nodes and set the vibration thresholds for them. Finally, based on the chosen sentry nodes, we carried out a test to validate the performance of the wakeup mechanism in terms of wakeup delay in H_1 and false positive ratio $\mathscr{P}^+$ in H_0.

In the heel drop test H_1, Fig. 3.15 illustrates one-second vibration data recorded at four nodes shown in Fig. 3.14. Obviously, node deployed near the event location (i.e., at 10.5th floor) has the highest vibration amplitude among the four nodes. In contrast, Fig. 3.16a illustrates the 8-min data measured at the node on 10th floor in H_0. The small peaks in the figure correspond to the vibration caused by the normal walk. Then using the peak-SVM method described in Sect. 3.6.2, the p_i^- and p_i^+ are illustrated in Fig. 3.16b. Different from the bridge example described in Sect. 3.6.2, due to the nature of this repeated heel drop test, δt_i is not determined by the structure

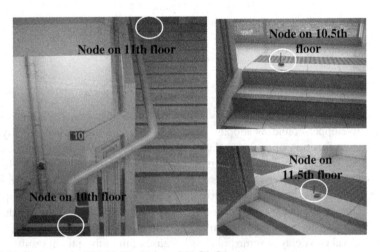

Fig. 3.14 Nodes on the 10th, 10.5th, 11th, and 11.5th floors

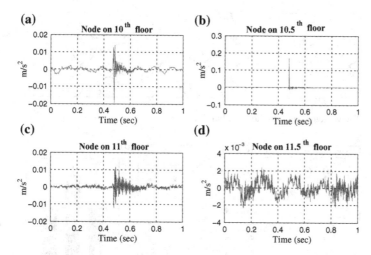

Fig. 3.15 The vibration data in a heel drop test by nodes deployed on the 10th, 10.5th, 11th, and 11.5th floors

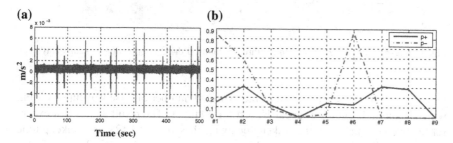

Fig. 3.16 **a** Data of node on the 10th floor when H_0. **b** The p_i^- and p_i^+ for each node

itself, but by the interval of heel drop test. Using the information, t_i can be calculated using Eq. 3.5.

The t_{ij} in the wakeup graph is determined through experiment. For each pair of nodes that can wake up each other, we carried out a total of 50 tests to find out t_{ij}. In these test, the clocks of the nodes are synchronized using FTSP protocol [11]. After outliers are deleted, we take the average time as the t_{ij} between node v_i and v_j. Figure 3.17 shows the wakeup graph in which the green-colored nodes at the bottom are the virtual nodes and the weight of each edge is illustrated.

We let $\alpha = 0.2$ in this test. Figure 3.18 shows the results of the three methods, respectively. The selected sentry nodes are the red-colored nodes at the bottom. It can be seen from Fig. 3.18d that among these methods, the R–K method again outperforms the other two methods.

Finally, we validate whether the sentry nodes selected by the R–K method actually meet the expected values. First, we repeat H_0 test and let the person walk as before. The test lasts for about 5 min, and we recorded 5 false wakeups, while $\alpha = 0.2$ with

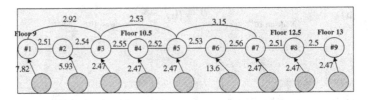

Fig. 3.17 The wakeup graph in the LSK building test

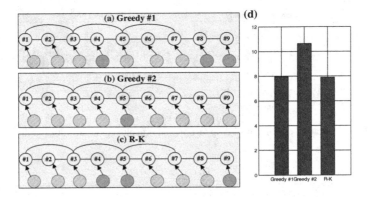

Fig. 3.18 **a–c** Sentry nodes chosen in the three methods, **d** the corresponding wakeup delays

time slot length $\Delta t = 20$ s indicates the expected number of false wakeups in 5 min which is about 3. Then we repeat H_1 test for 20 times. The result shows that except for the two tests in which the wakeup delay reaches 10.8 and 12 s, the wakeup time during the other tests ranges from 7.6−8.9 s, close to the theoretical value 7.91 s. The discrepancies in the number of false positive wakeups and the wakeup delay are mainly attributed to the errors in the drop-heel intervals, the statistical property contained in both training data and the validation data, and the possible difference contained between them. Nevertheless, through this experiment, this wakeup mechanism along with the sentry node placement method has shown great promise to realize fast, reliable, and network-wide wakeup in real applications.

References

1. J. Ansari, D. Pankin, P. Mähönen, Radio-triggered wake-ups with addressing capabilities for extremely low power sensor network applications. Int. J. Wirel. Inf. Netw. **16**(3), 118–130 (2009)
2. C.A. Balanis, *Antenna Theory: Analysis and Design/Constantine A. Balanis. J* (Wiley, New York, 1982)
3. M. Buettner, G.V. Yee, E. Anderson, R, Han. X-mac: a short preamble mac protocol for duty-cycled wireless sensor networks, in *SenSys2006* (2006)

4. K. Chebrolu, B. Raman, N. Mishra, P.K. Valiveti, R. Kumar, Brimon: a sensor network system for railway bridge monitoring, in *Proceeding of the 6th International Conference on Mobile Systems, Applications, and Services* (ACM, 2008), pp. 2–14
5. T.H. Cormen, *Introduction to Algorithms*, MIT Electrical Engineering and Computer Science Series (MIT Press, Cambridge, 2001)
6. P. Dutta, M. Grimmer, A. Arora, S. Bibyk, D. Culler, Design of a wireless sensor network platform for detecting rare, random, and ephemeral events, in *IPSN2005* (2005)
7. L. Gu, J.A. Stankovic, Radio-triggered wake-up for wireless sensor networks. Real-Time Syst. **29**(2), 157–182 (2005)
8. H. Kellerer et al., *Knapsack Problems* (Springer, Heidelberg, 2010)
9. X. Liu, S. Tang, J. Cao, Technical report: Enabling fast and reliable network-wide event-triggered wakeup in wireless sensor networks, http://imc.comp.polyu.edu.hk/isensnet/lib/exe/fetch.php?media=main2.pdf
10. G. Lu, D. De, M. Xu, W.Z. Song, B. Shirazi, A wake-on sensor network, in *Proceedings of the 7th ACM Conference on Embedded Networked Sensor Systems* (ACM, 2009), pp. 341–342
11. M. Maróti, B. Kusy, G. Simon, Á. Lédeczi, The flooding time synchronization protocol, in *Proceedings of the 2nd International Conference on ENSS* (2004), pp. 39–49
12. J.A. Rice, K. Mechitov, S.H. Sim, T. Nagayama, S. Jang, R. Kim, B.F. Spencer Jr, G. Agha, Y. Fujino, Flexible smart sensor framework for autonomous structural health monitoring. Smart Struct. Syst. **6**(5–6), 423–438 (2010)
13. L.C.L.P. Sub, 1ghz rf transceiver cc1101 datasheet. Texas Instruments (2012)
14. J.A.K. Suykens, J. Vandewalle, Least squares support vector machine classifiers. Neural Process. Lett. **9**(3), 293–300 (1999)
15. E.L. Wilson, A. Habibullah, Sap2000structural analysis users manual. Computers and Structures Inc (1998)

Chapter 4
Design of Distributed SHM Algorithms Within WSNs—A Cluster-Based Approach

4.1 Introduction

One of the most energy consuming operations in a WSN is wireless data transmission. To overcome this, computation power on the wireless sensor node can be utilized. Instead of streaming the sampled data directly to a central unit, the collected data are processed and only the processed information, which uses fewer bits than the original one, is transmitted. From this perspective, embedding SHM algorithms within a WSN can be an effective way to decrease the energy consumption.

Among a large variety of SHM algorithms, modal analysis has been widely used in SHM. From deployed sensors, the vibration characteristics (i.e., modal parameters) of the structures are identified using modal analysis. Modal parameters are the internal properties of a structure and can be used to detect and locate possible structural damage. In this chapter, we mainly focus on how to extract modal parameters in a WSN.

So far, WSN-based SHM systems which implement modal analysis generally have two types of architectures: centralized and distributed. A typical centralized architecture is illustrated in Fig. 4.1a. Systems of this architecture are intrinsically equivalent to wire-based counterpart [6]. In these systems, measurements collected by each sensor node are wirelessly transmitted to a central unit where modal analysis is implemented to obtain the modal parameters of the whole structure. Classic modal identification algorithms such as the eigen realization algorithm (ERA) [5], can be directly used without any modification. However, systems of this architecture have poor scalability and relatively high energy consumptions.

To address the aforementioned problems, civil researchers have proposed a method to implement modal analysis in a fully distributed way. The system architecture is illustrated in Fig. 4.1b. In this method, each sensor node implements modal analysis based on simple peak-picking technique [1]. The modal parameters identified from each sensor node are then transmitted to a central unit, where these modal parameters are assembled together. The particular advantage of these systems is

© The Author(s) 2016
J. Cao and X. Liu, *Wireless Sensor Networks for Structural Health Monitoring*,
SpringerBriefs in Electrical and Computer Engineering,
DOI 10.1007/978-3-319-29034-8_4

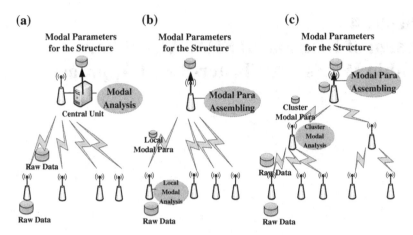

Fig. 4.1 Three WSN architectures for modal analysis, **a** centralized, **b** distributed, and **c** cluster-based

energy efficiency: local modal analysis is made without data exchange and transmitting only the modal parameters requires much less energy than streaming the raw data. However, since each sensor node in this approach performs modal analysis based on its own measured data without any collaboration with others, input change or measurement noise can easily degrade the identified local modal parameters and the errors cannot be reduced in the assembling process at the central unit.

In this chapter, to address the problems of both architectures above, we proposed a cluster-based approach for modal analysis in SHM. This architecture is presented in Fig. 4.1c. In this approach, the whole network is partitioned into a number of single-hop clusters. A cluster head (CH) is designated in each cluster to perform intracluster modal analysis using traditional centralized modal identification algorithms. The identified modal parameters in each cluster are then assembled together to obtain the modal parameters for the whole structure. Compared with the centralized approach, the cluster-based modal analysis limits the number of sensor nodes and hop count in each cluster, thus can be more energy efficient and scalable. Compared with the distributed approach, classic modal parameter identification techniques which use data-level fusion can be used in each cluster to obtain more reliable and accurate results. This cluster-based approach is therefore suitable for WSN-based SHM systems. In this approach, how to partition sensor nodes is critical. Clustering should address requirements from both modal analysis and wireless sensor networks (particularly in terms of energy). This optimal clustering problem is the focus of this chapter.

Strictly speaking, a work proposed in [10] can be regarded as a simplified cluster-based modal analysis where each cluster contains exact two sensor nodes. However, the authors did not consider other topologies and the problem of how to optimize clustering to decrease energy consumption of WSNs. Also, the chapter assumes all the sensor nodes are in a single communication range.

Table 4.1 Summary of notations used in this chapter

$\boldsymbol{\Psi}_k$	The kth mode shape vector of the structure
p	The number of mode shape vectors to be identified
$G_{xy}(\omega)$	The cross spectral density ($x \neq y$) and power spectral density ($x = y$)
N	The total data amount
M, c, n_i	The total number of sensor nodes, the number of generated clusters, and the number of sensor nodes in cluster S_i
n_t	Length of each section to calculate CSD
n_d	Number of averages
e_S, e_R, e_T	Energy consumed for sampling/rece./trans. one data
e_{Markov}, e_{ERA}	Energy consumed for calculating the Markov parameters and the remaining part of the ERA, resp.

Clustering has been exhaustively studied by researchers in computer science engineering and has been used in many applications to improve system performance, particularly in terms of local resources arbitration (e.g., in MAC protocol) and system scalability. Many clustering algorithms, such as LEACH [3], HEED [9], have been proposed. However, we will show in the following sections our clustering problem has some particular requirements from modal analysis and the existing clustering algorithms cannot be directly applied.

In the reminder of this chapter, we will first give the basic concept of modal parameters, then the techniques adopted for modal analysis and assembling method are described. Table 4.1 summarizes the notations used in this chapter.

4.2 Preliminary: Modal Parameters and the ERA

According to the vibration theory, every structure has tendency to oscillate with much larger amplitude at some frequencies than others. These frequencies are called natural frequencies. For a structure with n degrees of freedom, it has n natural frequencies. When a structure is vibrating under one of its natural frequencies, the corresponding vibration pattern it exhibits is called the mode shape for this natural frequency. For example, the natural frequencies and the corresponding mode shapes of a n-degrees of freedom structure are denoted, respectively, as

$$\Omega = \{f_1, f_2, \ldots, f_n\}, \quad \boldsymbol{\Psi} = [\boldsymbol{\Psi}_1, \boldsymbol{\Psi}_2, \ldots, \boldsymbol{\Psi}_n] \tag{4.1}$$

where f_i and $\boldsymbol{\Psi}_i = [\phi_i^1, \phi_i^2, \ldots, \phi_i^n]^T$ ($i = 1, \ldots, n$) are the ith natural frequency and the corresponding mode shape, respectively. $\phi_i^k (k = 1, 2, \ldots, n)$ is the value of $\boldsymbol{\Psi}_i$ at the kth degree of freedom. For convenience, ω_i and $\boldsymbol{\Psi}_i$ are also called **modal parameters** corresponding to the ith **mode** of a structure. As an example, Fig. 4.2

(a)   **(b)**   **(c)**

Fig. 4.2 The first three modal parameters of a cantilever beam. **a** 1st mode shape, 1st natural frequency $= 0.9\,$Hz, **b** 2nd mode shape, 2nd natural frequency $= 5.7\,$Hz, **c** 3rd mode shape, 3rd natural frequency $= 15\,$Hz

shows the first few natural frequencies and mode shapes of a typical cantilevered beam.

In the reminder of this section, we briefly introduce a classical modal parameter identification method, the ERA.

Assume a total of m sensors are deployed and the collected data are denoted as a data matrix $\mathbf{X}$:

$$\mathbf{X} = \begin{bmatrix} x^1(1) & x^2(1) & \dots & x^m(1) \\ x^1(2) & x^2(2) & \dots & x^m(2) \\ & & \dots & \\ x^1(N_{ori}) & x^2(N_{ori}) & \dots & x^m(N_{ori}) \end{bmatrix} = [\mathbf{x}^1, \dots \mathbf{x}^m] \qquad (4.2)$$

where $x^i(k)$ is the data sampled by the ith sensor at kth time step and N_{ori} is the total number of data points collected in each node. The ith column of $\mathbf{X}$, $\mathbf{x}^i$, represents the sequence of data points measured from the ith sensor.

Given the collected data in Eq. 4.2, the cross spectral density (CSD) between each of the signal $\mathbf{x}^i$ and a reference signal $\mathbf{x}^{ref}$ is calculated using the Welch's method [2]. Generally speaking, measured signal from any sensor node can be selected as $\mathbf{x}^{ref}$. For two signals $\mathbf{x}$ and $\mathbf{y}$, the Welch method first divides $\mathbf{x}$ and $\mathbf{y}$ into n_d number of overlapping segments. The CSD of $\mathbf{x}$ and $\mathbf{y}$, denoted as G_{xy}, is then calculated as

$$G_{xy}(\omega) = \frac{1}{n_d \cdot N} \sum_{i=1}^{n_d} X_i^*(\omega) \cdot Y_i(\omega) \qquad (4.3)$$

where $X_i(\omega)$ and $Y_i(\omega)$ are the Fourier transforms of the ith segment of $\mathbf{x}$ and $\mathbf{y}$, and '*' denotes the complex conjugate. N is data points in each segment of $\mathbf{x}$ (or $\mathbf{y}$) as well as the obtained $G_{xy}(\omega)$. N is generally taken as a power of two value such as 1024 or 4096 to give reasonable results. To decrease the noise, n_d practically ranges from 10 to 20. Note that if 50 % overlap of two consecutive segments is employed in the CSD estimation, $\mathbf{x}$ and $\mathbf{y}$ need to contain at least $N/2 * (n_d + 1)$ data points. For convenience, unless specified, we set $N = 4096$, $n_d = 20$ and the overlap is 50 % in this chapter. Correspondingly, the number of raw data to be sampled at each sensor node should be at least 43,008.

The CSD of $\mathbf{x}^{ref}$ with each signal in $\mathbf{X}$ can be written as a matrix:

$$\mathbf{CSD}(\omega) = [G_{x^1 x^{ref}}(\omega), \dots, G_{x^m x^{ref}}(\omega)]^T \qquad (4.4)$$

Having obtained the $\mathbf{CSD}(\omega)$, the inverse Fourier transform (IFFT) is carried out on all the CSDs to obtain a cross-correlation function (CCF) matrix:

$$\mathbf{CCF} = [CCF_{\mathbf{x}^1\mathbf{x}^{ref}}, \ldots, CCF_{\mathbf{x}^m\mathbf{x}^{ref}}]^T \tag{4.5}$$

where $CCF_{\mathbf{x}^i\mathbf{x}^{ref}}$ is the CCF between $\mathbf{x}^i$ and $\mathbf{x}^{ref}$, and is calculated as the IFFT of $G_{\mathbf{x}^i\mathbf{x}^{ref}}(\omega)$. From the $\mathbf{CCF}$, we can find out a series of parameters $\{\mathbf{Y}(1), \mathbf{Y}(2), \ldots\}$ which are called as **Markov parameters**. At k time step, the Markov parameter $\mathbf{Y}(k)$ is defined as:

$$\mathbf{Y}(k) = [CCF_{\mathbf{x}^1\mathbf{x}^{ref}}(k), \ldots, CCF_{\mathbf{x}^m\mathbf{x}^{ref}}(k)]^T \tag{4.6}$$

Note that due to the symmetric property of the CCF, only first half of the obtained CCF are used as the Markov parameters (i.e., $k = 1, \ldots, N/2$).

With the obtained $\mathbf{Y}(1), \mathbf{Y}(2), \ldots, \mathbf{Y}(N/2)$, the ERA begins by forming a Hankel matrix. Moreover, instead in the original form of Hankel matrix, the ERA algorithm allows us to use a modified one shown below:

$$\mathbf{H}(k-1) = \begin{bmatrix} \mathbf{Y}(k) & \mathbf{Y}(k+\alpha) & \cdots & \mathbf{Y}(k+\beta\alpha-\alpha) \\ \mathbf{Y}(k+1) & \mathbf{Y}(k+\alpha+1) & \cdots & \\ \vdots & & & \\ \mathbf{Y}(k+\alpha-1) & \mathbf{Y}(k+2\alpha-1) & \cdots & \mathbf{Y}(k+\beta\alpha-1) \end{bmatrix} \tag{4.7}$$

Note that we only need to calculate two Hankel matrices $\mathbf{H}(0)$ and $\mathbf{H}(1)$ $(\mathbf{H}(0), \mathbf{H}(1) \in \mathbb{R}^{\alpha m \times \beta})$ to identify modal parameters. α and β in Eq. 4.7 correspond to the number of block rows and columns in the Hankel matrix. In this chapter, $\alpha = 20$ and $\beta = 100$, which are large enough to accurately identify the modal parameters of important modes [5].

The ERA then performs the SVD on $\mathbf{H}(0)$:

$$\mathbf{H}(0) \overset{svd}{=} \mathbf{USV}^T \tag{4.8}$$

Both $\mathbf{U}$ and $\mathbf{V}$ are unitary matrices. In addition, according to the vibration theory [4], for a structure whose vibration is dominated by the first n modes, the rank of $\mathbf{H}(0)$ is only $2n$ and the singular values in $\mathbf{S}$ has $2n$ nonzero values: $\mathbf{S} = \begin{bmatrix} \mathbf{S}_{2n} & \mathbf{0} \\ \mathbf{0} & \mathbf{0} \end{bmatrix}$, where $\mathbf{S}_{2n} = diag(d_1, \ldots, d_{2n})$, $d_1 \geq \cdots \geq d_{2n} > 0$ are the singular values of $\mathbf{H}(0)$. Correspondingly, $\mathbf{H}(0)$ can be re-expressed as:

$$\mathbf{H}(0) = \mathbf{U}_{2n}\mathbf{S}_{2n}\mathbf{V}_{2n}^T \tag{4.9}$$

where $\mathbf{U}_{2n} \in \mathbb{R}^{\alpha m \times 2n}$ and $\mathbf{V}_{2n} \in \mathbb{R}^{\beta \times 2n}$ are the first $2n$ columns of the matrices $\mathbf{U}$ and $\mathbf{V}$, respectively. Notice that although $\mathbf{U}$ and $\mathbf{V}$ are not unitary, they still contain orthogonal columns: (i.e., $\mathbf{U}_{2n}^T\mathbf{U}_{2n} = \mathbf{V}_{2n}^T\mathbf{V}_{2n} = \mathbf{I}_{2n}$).

Using Eq. 4.9 and the Hankel matrix $\mathbf{H}(1)$ constructed from Eq. 4.7, two matrices $\mathbf{A}$ and $\boldsymbol{\Gamma}$ are found [5]:

$$\mathbf{A} = \mathbf{S}_{2n}^{-1/2}\mathbf{U}_{2n}^{T}\mathbf{H}(1)\mathbf{V}_{2n}\mathbf{S}_{2n}^{-1/2}, \; \boldsymbol{\Gamma} = [\mathbf{I}_{2n}, \mathbf{0}]\mathbf{U}_{2n}\mathbf{S}_{2n}^{-1/2} \quad (4.10)$$

Performing the eigen decomposition on $\mathbf{A}$: $\mathbf{A} = \boldsymbol{\Phi}\Lambda\boldsymbol{\Phi}^{-1}$, The mode shape $\boldsymbol{\Psi}$ at the measured degrees of freedom, denoted as $\hat{\boldsymbol{\Psi}}$ can be calculated as $\hat{\boldsymbol{\Psi}} = \boldsymbol{\Gamma}\boldsymbol{\Phi}$.

It can be seen that the ERA can be largely divided into two stages. In the first stage, the Markov parameters are identified. These Markov parameters are then used to identify the modal parameters in the second stage.

4.3 Cluster-Based Modal Analysis and Its Energy Consumption

In the cluster-based modal analysis, deployed sensor nodes are partitioned into a number of single-hop clusters and each CH performs intracluster modal analysis to extract local mode shapes. Since mode shapes of a cluster only contain elements corresponding to the sensor nodes in that cluster, the mode shapes in all the clusters need to be assembled to obtain the mode shapes defined on all the deployed sensor nodes. The whole process is illustrated in Fig. 4.3.

In each cluster, the ERA is used first to calculate the cross spectral density (CSD) between the CH and each of the cluster member as in Eq. 4.4. Having obtained the CSD, the inverse Fourier transform (IFFT) is carried out on all the CSDs to obtain a cross-correlation function (CCF) matrix as in Eq. 4.5.

Traditionally, CH collects the raw data from all its cluster members, calculates CSDs, and then uses the ERA to identify mode shapes. However, the model-based data aggregation method proposed by [7] can be used here to decrease the energy consumption. In this approach, instead of collecting measurements data from cluster members, CH broadcasts its time record of length n_t. On receiving the record, each cluster member calculates its CSD and stores it locally. This procedure will

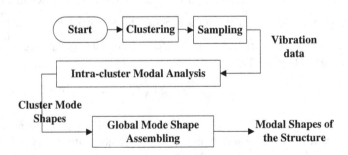

Fig. 4.3 Overview of cluster-based modal analysis process

be repeated n_d times, until the CSD is obtained according to Eq. 4.3. Each cluster member then transmits the first half of the corresponding CCF to the CH.

Based on the discussion above, we can estimate the energy consumption of intracluster modal analysis. To obtain the mode shapes of a cluster S_i, the total energy consumption in S_i, denoted as $cost(S_i)$, consists of the following three parts:

$$cost(S_i) = Er_s(S_i) + Er_c(S_i) + Er_a(S_i) \qquad (4.11a)$$

where $Er_s(S_i)$, $Er_c(S_i)$ and $Er_a(S_i)$ are the energy consumed in data sampling, intracluster wireless communication and computation associated with modal analysis, respectively.

Assume a cluster S_i contains a total of n_i sensor nodes, then sampling cost $Er_s(S_i)$ is

$$Er_s(S_i) = n_i \cdot N \cdot e_S \qquad (4.11b)$$

where N is the total amount of time history record sampled in each sensor. Assuming 50 % overlapping, $N = (n_d/2 + 1/2)n_t$. e_S is the energy for sampling one data. We assume that n_d, n_t, N and e_S are fixed in this chapter.

The intracluster wireless communication cost $Er_c(S_i)$ is

$$Er_c(S_i) = N \cdot e_T + (n_i - 1)N \cdot e_R + (n_i - 1)\frac{n_t}{2}(e_T + e_R) \qquad (4.11c)$$

where e_T and e_R are the energy cost for transmitting and receiving one data, respectively. The first two terms at the right side of Eq. 4.11c are the energy consumed when CH broadcasts its time history data and when all the cluster members receive the broadcasts, respectively. The last term is the energy consumption when the $(n_i - 1)$ cluster members transmit back their correlation functions to the CH.

The computation cost $Er_a(S_i)$ can be formulated as

$$Er_a(S_i) = n_i \cdot e_{Markov} + e_{ERA}(n_i) \qquad (4.11d)$$

where e_{Markov} is the energy consumed for each node before obtaining the Markov parameters, and e_{ERA} is the energy used in CH when it carries out the remaining steps of the ERA for mode shape identification. e_{Markov} is fixed given n_t and n_d. e_{ERA} is dependent on n_i and number of mode shape vectors p to be identified. Given p, $e_{ERA}(n_i)$ is not a linear function of n_i since the ERA involves complex matrix computations including SVD and matrix inversion. This point is demonstrated in Fig. 4.4, where the computation time of our SHM mote to implement the ERA for different cluster sizes is illustrated. The fitting function is also illustrated in the figure. It can be seen that with the increase of n_i, the time consumed, which is the indicator of energy consumption, is quadratically increased.

From the equations above, we have $cost(S_i) = cost(n_i)$, indicating that the energy consumption of a cluster is only associated with the number of nodes in this cluster. It is of interest to see that if possible, whether to generate small-sized clusters or large-

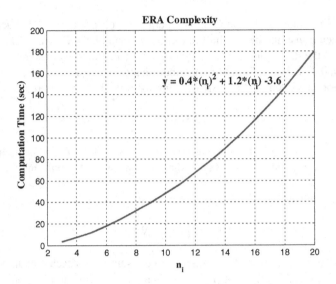

Fig. 4.4 The complexity of the ERA

sized clusters is more energy efficient. To find the answer, we assume M sensor nodes can be partitioned into equal-sized clusters of size n, then the number of clusters $c = M/n$. The optimal cluster size, denoted as n_{opt}, can be obtained by looking for the n that minimizes the average energy consumption per node defined as:

$$Epn(n) = \frac{c \cdot cost(n)}{M} = N(e_S + \beta) + e_{Markov} + \frac{N(e_T - \beta)}{n} + \frac{e_{ERA}(n)}{n} \quad (4.12)$$

where $\beta = e_R + \frac{n_i}{2N}(e_T + e_R)$.

The 3rd term in the right side of the Eq. 4.12 indicates that in terms of wireless communication, partitioning sensor network into large-sized clusters is preferred when $e_T \geq \beta$ while generating small-sized clusters is better if otherwise. The 4th term tells us that small cluster size n is more energy efficient in terms of computation considering that $e_{ERA}(n)$ is a quadratic function of n. As a result, there does not exist a rule of thumb for clustering and we have different optimal cluster sizes for different conditions. As an example, some parameters obtained by some real tests of our SHM Mote are listed in Table 4.2. Based on Table 4.2, Fig. 4.5a shows various optimal cluster sizes, illustrated as red dots in the figure, when the transmission power e_T is set to be from $e_T = e_R$ to $e_T = 5e_R$. It can be seen that when $e_T = e_R$, the smaller the cluster size, the better. (Note that the ERA requires that the number of sensor nodes in each cluster should be at least larger than p). With the increase of e_T, the optimal cluster size is increased but does not go unbounded considering the energy consumption of the ERA for large-sized clusters.

Clustering using minimum dominating set [8] cannot be directly applied to solve our clustering problem since it mainly aims to find as small number of clusters as

Table 4.2 Parameters used in Fig. 4.5

N	p	n_t	n_d	e_S (mAh)	e_R (mAh)	e_{Markov} (mAh)	e_{ERA} (mAh)
10752	3	1024	20	$1.1e{-}4$	$5e{-}4$	0.5	$0.0417(0.4n_i^3 + 1.2n_i - 3.6)$

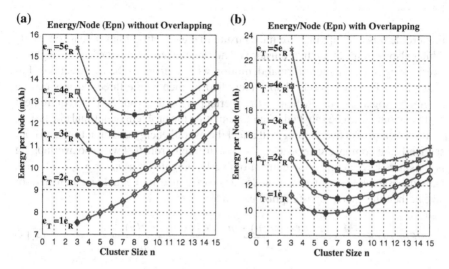

Fig. 4.5 The optimal cluster sizes in different conditions. **a** No overlapping. **b** With overlapping

possible. Also, in the discussion so far, we assume that no overlapping nodes exist in the clusters. However, we will show in the following section that a necessary condition for cluster-based modal analysis is that all the generated clusters must be connected through the overlapping nodes. This requirement further increases the difficulty of the clustering problem.

4.4 Determination of Global Mode Shapes Through Mode Shape Assembling

After the mode shapes in all clusters have been identified, they need to be stitched together to obtain mode shapes defined on all of deployed sensor nodes.

However, since mode shape vectors identified in a cluster only represent the relative vibration amplitudes at cluster sensor nodes, mode shapes of different clusters may not be able to be assembled together. This can be demonstrated in Fig. 4.6a, where the deployed 12 sensor nodes are partitioned into three clusters to identify the 3rd mode shape. Although the mode shape of each cluster is correctly identified, we still cannot obtain the mode shapes for the whole structure. The key to solve

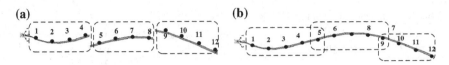

Fig. 4.6 Mode shape assembling when **a** clusters do not overlap, **b** clusters overlap

this problem is overlapping. We must ensure that each cluster has at least one node which also belongs to another cluster and all the clusters are connected through the overlapping nodes (a more formal definition will be given in the next section). For example, in Fig. 4.6b, mode shapes identified in each of the three clusters can be assembled together with the help of the overlapping nodes 5 and 9. This requirement of overlapping must be satisfied when formulating the problem of optimal clustering.

It is obvious that overlapping will affect the overall energy consumption, and consequently the optimal cluster size n_{opt} will be different from that when no overlapping is considered. By defining the number of overlapping nodes as $n_o = \sum_{i=1}^{c} |S_i| - M$, and still assume these M sensor nodes are partitioned into equal-sized clusters of size n, then the energy consumption per node becomes

$$Epn'(n) = \frac{(M + n_o)/n \cdot cost(n) - n_o \cdot N \cdot e_S}{M} = \frac{cost(n)}{n} + \frac{n_o}{M} \cdot \kappa, \qquad (4.13)$$

where $\kappa = N \cdot \beta + e_{NEXT} + \frac{N(e_T - \beta)}{n} + \frac{e_{ERA}(n)}{n}$. Considering the fact that unnecessary overlapping will cause extra energy consumption and the number of overlapping nodes should be kept as small as possible, we require that $n_o \geq \frac{M + n_o}{n} - 1$. Therefore,

$$Epn'(n) \geq \frac{cost(n)}{n} + \frac{1 - n/M}{n - 1} \kappa \qquad (4.14)$$

The right side of Eq. 4.14 essentially provides a lower bound of energy consumption that clustering can achieve when the overlapping constraint is considered. The optimal cluster size n_{opt} can be calculated by minimizing n in Eq. 4.14. For example, n_{opt} for the parameters in Table 4.2 are illustrated in Fig. 4.5b.

By comparing Fig. 4.5a with Fig. 4.5b, it also can be easily seen that optimal cluster size is larger when overlapping constraint is considered. Clustering which generates small-sized clusters may not be energy efficient since a large number of overlapping nodes can cause extra energy consumption in terms of communication and computation.

Also should be noted is that the optimal cluster size n_{opt}, either obtained by Eq. 4.12 or by Eq. 4.14, is not affected by actual network topology. In a dense network, it is more possible to achieve the obtained optimal cluster size, and therefore the total energy will be lower than a sparse network.

We do not consider the inter-cluster communication in this chapter simply because delivering obtained mode shapes requires significantly less energy than other processes.

4.5 Optimal Clustering

In this section, we will formulate the optimal clustering problem. The objective of clustering is that the generated clusters can minimize the energy consumption of overall modal analysis. Clustering also has to satisfy the following constraints (1) each sensor node belongs to at least one of the generated clusters, (2) sensor nodes in each cluster is within a single communication range to its CH, (3) number of sensor nodes in each cluster is larger than p (p: the number of mode shape vectors to be identified) (4) all the clusters are connected together through the overlapping nodes. More formally, problem is formulated as follows: Given a sensor network $G = (V, E)$, find a clustering scheme that can cluster these V sensor nodes into a set of clusters, denoted as $C = \{S_1, S_2, S_3, \ldots\}$, subject to the following constraints:

1. $\bigcup_{S_i \in C} = V$
2. Let the sub-graph for S_i is $G(S_i, E_i)$, where $E_i \subseteq E$. Then $\forall S_i \in C$, $\exists s_i \in S_i$, such that there is an edge $a_{ij} \in E_i$ between s_i and any other $s_j \in S_i (s_i \neq s_j)$
3. $\forall S_i \in C, |S_i| \geq p$
4. $\forall S_i, \exists S_j \in C, (i \neq j), S_i \cap S_j \neq \emptyset$
5. $\forall C' \subseteq C, (\bigcup_{S_i \in C'} S_i) \cap (\bigcup_{S_j \in C-C'} S_j) \neq \emptyset$

Objective

- Minimize $\sum_{S_i \in C} cost(S_i)$

The first constraint is set because we wish to find the mode shapes defined on all the deployed sensor nodes. The second constraint is to ensure only single-hop clusters are generated. Constraint 3 is required by the ERA algorithm. Constraints 4 and 5 are used to describe that generated clusters are overlapping and connected.

The above clustering problem is an optimization problem. We will prove that the decision version of the problem is NP-complete which is defined as *given a threshold* k, *does there exist a cluster set* $C = \{S_1, S_2, S_3, \ldots\}$, *which satisfy all the constraints above and whose total energy cost* $\sum_{S_i \in C} cost(S_i)$ *is equal of smaller than k?* We will prove that this decision version is NP-complete.

Proof It is easy to find out this clustering problem is NP. Given a cluster set C, all the constraints above, include constraints 4 and 5 can be checked in a polynomial time. The detailed proof of this part is omitted for brevity.

We show this decision version of the clustering problem is NP-hard by reducing the set cover problem to it. The set cover problem is defined as follows.

Given

1. A universe V'
2. A set of $S' = \{S'_1, S'_2, \ldots\} \subseteq V'$
3. The cost function for each subset $S'_i \in S'$: $cost'(S'_i)$
4. A number k'.

Find: If there is a subset $C' \subseteq S'$ which satisfies

1. $\bigcup\limits_{S'_i \in C'} S'_i = V'$

2. $\sum\limits_{S'_i \in C'} cost'(S'_i) \leq k'$.

To reduce the set cover problem to the clustering problem, we construct a sensor network $G = (V, E)$ from the inputs of set cover problem in the following way: The vertices $V = V' \bigcup X$, where $X = \{x_1, x_2, \ldots x_p\}$ is a set of p virtual nodes. To construct the edges E, for each $S'_i \in S'$, we first choose an arbitrary node $s'_i \in S'_i$, then we add an edge between s'_i and any other node $s'_j \in S'_i (s'_j \neq s'_i)$. We also add an edge between s'_i and any virtual node in X. The cost function in the clustering problem $cost(\cdot) = cost'(\cdot)$. We also define that by adding/deleting any virtual node x to/from any group will not affect the cost function. The energy threshold $k = k'$.

With this transformation, it can be easily proved that (1) Assume $C' = \{S'_1, S'_2, \ldots\}$ is a solution to the set cover problem, then $C = \{S_1, S_2, \ldots\}$ is a solution to the clustering problem, where $S_i = S'_i \bigcup X$. (2) Assume $G = (V, E)$ is constructed from the set cover problem and we have a solution $C = \{S_1, S_2, \ldots\}$ to the clustering problem, then $C' = \{S'_1, S'_2, \ldots\}$ is a solution to the set cover problem, where $S'_i = S_i - X$. The detailed proof is omitted for brevity.

By reducing the NP-complete set problem to our clustering problem, we have demonstrated that the decision version of our problem is NP-complete. Obviously, the original clustering problem is also NP-complete.

4.5.1 Proposed Methods for Energy Efficient Clustering

Two centralized algorithms are proposed to solve our clustering problem. These two algorithms use the similar idea of the greedy algorithm for the set cover problem but adopt different approaches to handle the extra constraints of clustering. In both of the algorithms, a set of candidate single-hop clusters is first established given the network $G = (V, E)$. Then the most cost-effective cluster is selected from this set, one at a time, until all the sensor nodes in V have been covered.

In the first algorithm, to find a candidate cluster set U, we first calculate the optimal cluster size n_{opt} according to Eq. 4.14. Then based on n_{opt}, one-hop neighbors of each node in V are partitioned. For each node $s_i \in V$, assume the one-hop neighbor set is Ne_{s_i}, if $|Ne_{s_i}| \geq n_{opt} - 1$, then each cluster in the cluster set contains a common element s_i and the remaining elements are the combinations of nodes in Ne_{s_i} with the length of $n_{opt} - 1$. When $|Ne_{s_i}| < n_{opt} - 1$, $C_i = \{s_i\} \bigcup Ne_{s_i}$. Note that we assume the network is dense enough such that each sensor node has at least p one-hop neighbors. The obtained cluster sets for all the nodes in V are combined together to obtain the candidate cluster set U.

The algorithm then selects the most cost-effective cluster $S_i \in U$, one at a time, until all the sensor nodes in V have been covered. The cost-effectiveness, denoted as λ, is defined as $\lambda = \frac{1}{|S_i \cup C_a| - |C_a|}$, where C_a represents the set of nodes covered so far. When selecting the most cost-effective cluster, we choose from the clusters in U which overlap with C_a. This strategy can ensure that all the selected sensor nodes will be connected through the overlapping nodes. If more than one candidate clusters which overlap with C_a have the same λ, the one which maximizes the total degrees of the remaining uncovered nodes (i.e., $U - S_i \bigcup C_a$) will be chosen. It can be seen that this algorithm divides the sensor nodes in V into as many single-hop clusters of size n_{opt} as possible while keeps the number of overlapping nodes into minimums (from $\lambda = \frac{1}{|S_i \cup C_a| - |C_a|}$, penalty is given to cluster having large number of overlapping nodes with C_a). Both of these two points are of importance to minimize the overall energy cost. The algorithm is shown as Algorithm 2.

The second algorithm uses different strategy to handle overlapping. First, the optimal cluster size n_{opt} is calculated based on Eq. 4.13 without considering overlapping constraint. When selecting the most cost-effective cluster, it is chosen from all the candidate clusters in U. Since the overlapping constraint is not considered when selecting cluster, after all the sensor nodes in V have been covered, the algorithm will test if all the clusters are connected through the overlapping nodes and add extra clusters to connect them if necessary. The basic idea is to identify all the isolated cluster groups and then find clusters to connect them. The detailed description is omitted for brevity. This algorithm is shown as Algorithm 3.

Based on our first centralized algorithm, we propose a distributed solution. In this solution, each node only needs its one-hop neighbors' information and communicates only with its one-hop neighbors. The clustering will start at a single controller node which usually is the sink node of the network.

Similar to Algorithm 2, each newly created cluster will be connected to at least one of the existing clusters. In the distributed algorithm, each node will maintain two lists of neighbors: unclustered and clustered. The lists will be sorted according to each neighbor's own number of unclustered neighbors. The nodes with fewer unclustered neighbors will do clustering or join other clusters first. This is done by assigning each node's execution of the algorithm to a specific time slot.

Each node has only three roles during clustering: unclustered, CM (cluster member), and CH. We illustrate the pseudocode based on the three roles in Algorithm 4.

Algorithm 2 Centralized algorithm 1

Require: $G = (V, E)$ and parameters listed in Table 4.2
 1: find n_{opt} which minimizes Eq. 4.13
 2: $U \leftarrow \emptyset\ C_a \leftarrow \emptyset$
 3: **for all** $n_i \in V$ **do**
 4:     $S_i \leftarrow \emptyset$
 5:     **for all** one-hop neighbor n_j of n_i **do**
 6:         $S_i = S_i \bigcup \{n_j\}$
 7:     **end for**
 8:     construct a cluster set C_i whose elements are the combinations taken of the nodes in S_i of length $n_{opt} - 1$.
 9:     $U = U \bigcup C_i$
10: **end for**
11: **repeat**
12:     $C_{cand} = $ all the clusters in U which overlap with C_a
13:     find a cluster S_i in C_{cand} with the smallest $\frac{1}{|S_i \cup C_a| - |C_a|}$
14:     $C_a = C_a \bigcup S_i$
15: **until** C_a covers V
Ensure: C_a

Algorithm 3 Centralized algorithm 2

Require: $G = (V, E)$ and parameters listed in Table 4.2
 1: find n_{opt} which minimizes Eq. 4.12
 2: The same with the 2 to 16 lines of Algorithm 2
 3: **repeat**
 4:     find a cluster S_i in U with the smallest $\frac{1}{|S_i \cup C_a| - |C_a|}$
 5:     $C_a = C_a \bigcup S_i$
 6: **until** C_a covers V
 7: Identify Isolated cluster groups (ICGs) in C_a
 8: Construct a graph $G_{ICG} = (V_{ICG}, E_{ICG})$:
 9: Run MST algorithm on G_{ICG} and get T
10: **for all** edges in T **do**
11:     Create an extra cluster C_e and add it to C_a
12: **end for**
Ensure: C_a

Each node will not execute the algorithm until the start of its own time slot. The input p is the minimum cluster size constraint and n_{opt} is the calculated optimal cluster size. Once a CH decides to choose certain node as CM, it will send a message req_{ch} and the corresponding CM will send $acpt_{ch}$ to acknowledge the selection. After the execution of the algorithm on a node, an unclustered node will become CH. At the end of the algorithm, each node will merge all its neighbors into one sorted list and assign time slots. The duration of the time slot is large enough so that a CH can perform the selection. $t[i]$ is the ith time slot from the end of the current time slot. It can be easily seen that the generated clusters from the distributed algorithm can satisfy all the constraints.

Algorithm 4 Distributed algorithm

Require: n_{opt}, p, unclustered neighbors un, clustered neighbors cn (un, and cn are both sorted in
  increasing order according to the number of unclustered neighbors)
1: **if** self is unclustered **then**
2:    Self becomes CH
3: **end if**
4: **if** self is CH **then**
5:    Select one node from cn as its member
6:    **if** $size(un) \geq n_{opt}$ **then**
7:       Construct the cluster as size of n_{opt} by selecting first n_{opt} nodes from un as CM
8:    **else if** $n_{opt} > size(un) \geq p$ **then**
9:       Construct a cluster by selecting all nodes in un as CM
10:   **else**
11:      First construct a cluster by selecting all nodes in un as CM then select more nodes from
         cn as CM until $cluster\_size = p$
12:   **end if**
13: **else if** self is CM **then**
14:   Broadcast a message to all neighbors saying the status is currently CM.
15: **end if**
16: $an = merge(un, cn)$
17: **for all** $n \in an$ that have not been assigned a time slot **do**
18:   Assign time slot $t[i]$ to n where i is the index of n in an
19: **end for**

4.6 Performance Evaluation

4.6.1 Simulation

We first use simulation data to demonstrate the effectiveness of our cluster-based
modal analysis approach and the clustering algorithms. We fix all associated parameters except two: the network topology and the transmission power e_T. Although
in some circumstances, these two factors are correlated, they are independently considered in this simulation. A total of 40 sensor nodes are randomly deployed starting
from a relatively sparse wireless sensor network with network degree, defined using
the node with the minimum degree in the network, is 3. Under this network degree,
we considered energy consumptions when the transmission power e_T is changing
from $5e{-}4$ to $2.5e{-}3$, with the interval of $5e{-}4$. The energy consumption of the
following four approaches are considered: (1) the traditional approach when all the
raw data are streamed back to a sink node, (2) the cluster-based modal analysis with
clusters generated from the 1st centralized algorithm, (3) from the 2nd centralized
algorithm, and (4) from the distributed algorithm. In the first approach, the sink node
is chosen to be the node whose short path tree (SPT) is the shortest among all the
other nodes. Also, no computation energy is included in this approach. A total of 500
simulations are performed and the average energy consumption of each approach is
calculated. The above procedure is repeated when the network degree is changed to
be 4 and 5.

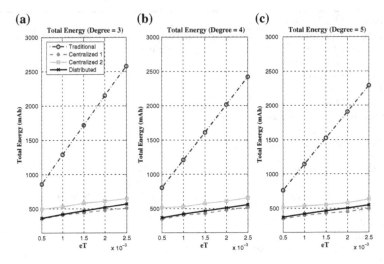

Fig. 4.7 The total energy in different scenarios when **a** network degree = 3, **b** network degree = 4, and **c** network degree = 5

The parameters associated with the simulation are listed in Table 4.2 and the simulation results are shown in Fig. 4.7. It can be seen that for the first traditional approach, the total energy is linearly increased with the increase of e_T and the slope is the length of the SPT rooted at the sink node. The cluster-based approach, either using clusters generated from centralized or distributed algorithms, is much more energy efficient compared with the traditional approach. This conclusion is more evident when the transmission power e_T is large. When $e_T = 2.5e-3$, the energy consumption of the cluster-based approach is about one-fifth of the traditional approach. For the cluster-based approach, it seems that using clusters generated from the 1st centralized algorithm and from the distributed algorithm are slightly better than the 2nd centralized algorithm.

To further demonstrate the importance of using optimal cluster size, we illustrate in Fig. 4.8 the energy consumption when the 1st algorithm chooses three different cluster sizes: $n_{opt} - 1$, n_{opt} and $n_{opt} + 1$. It can be seen that compared with other cluster sizes, clustering using designed optimal cluster size can achieve lower energy consumption.

4.6.2 Experiment

We have tested our cluster-based modal analysis approach through a real implementation. The wireless sensor nodes adopted are called SHM Mote and are particularly developed for general SHM applications (see Fig. 4.9a). A SHM Mote includes an

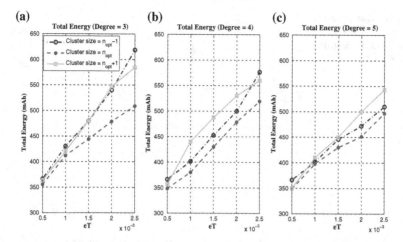

Fig. 4.8 The energy consumption using different cluster sizes when **a** network degree = 3, **b** network degree = 4, and **c** network degree = 5

Fig. 4.9 SHM Mote and testing structure. **a** SHM Mote. **b** Testing structure. **c** Network topology

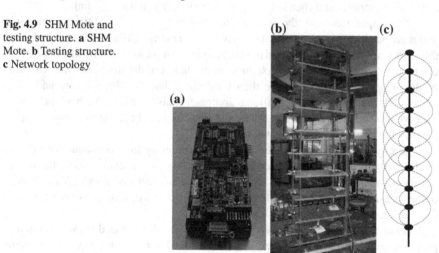

Intel Imote2, a sensor board, an external 32Mb non-volatile memory chip, an AM radio receiver for synchronized sensing, and a RF amplifier.

Figure 4.9b shows the setting of the lab test. The test building has 10 floors, at each floor, a Mote is deployed to monitor the structure's horizontal accelerations. We adjust the transmission power to be $e_T = 1e-5$ in this test after a few tests on link quality. Under this transmission power, the topology of the network along with the simplified structure is illustrated in Fig. 4.9c. We use a gateway node which is connected a computer for the control purpose, while this gateway can be removed in future implementation. The SHM Motes run modified TinyOS and are configured to sample the accelerometers in a synchronized manner at frequency of 512 Hz. Under

Fig. 4.10 Testing results.
a Clustering using
Algorithm 1. **b** Identified
Mode shapes

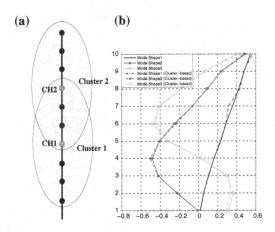

the command of gateway, each Mote starts collecting $N = 10{,}752$ data samples synchronously. The cluster information is calculated by the 1st centralized algorithm run in the computer and then is feed to each node through wireless link.

In this implementation, the optimal cluster size is $n_{opt} = 7$ and these ten sensor nodes are partitioned into two clusters as illustrated in Fig. 4.10a. Once each node has this information, the cluster-based modal analysis is implemented in each cluster, one cluster at a time, to identify the first three mode shapes of the structure. The obtained mode shapes of each cluster are then transmitted back to the gateway and then assembled. For comparison, traditional approach is also used in which all measured data are transmitted the gateway where the ERA using all the measurement data to identify the mode shapes of the whole structure.

Figure 4.10b illustrates the identified mode shapes by the traditional centralized modal analysis and by our cluster-based modal analysis. It can be seen that using cluster-based modal analysis, the mode shapes can be identified without losing much of the accuracy. Moreover, we tested that the energy communication cost is decreased from 259 to 152 mAh.

Our experiment shows that even for a small-scaled WSN-based SHM system, the proposed approach can reduce energy by 40 %. In the future, we plan to use our system to estimate mode shapes of large civil structures. With the increase of transmission power, it can be expected that advantage of this approach is more significant in these applications.

References

1. D.J. Ewins, *Modal Testing: Theory and Practice* (Research Studies, England, 1984)
2. F.J. Harris, On the use of windows for harmonic analysis with the discrete fourier transform. Proc. IEEE **66**(1), 51–83 (1978)
3. W.B. Heinzelman, A.P. Chandrakasan, H. Balakrishnan, An application-specific protocol architecture for wireless microsensor networks. IEEE Trans. Wirel. Commun. **40**(8), 660–670 (2002)

4. D.J. Inman, *Engineering Vibrations* (2006)
5. J.N. Juang, R.S. Pappa, Eigensystem realization algorithm for modal parameter identification and model reduction. J. Guid. Control Dyn. **8**(5), 620–627 (1985)
6. S. Kim, S. Pakzad, Health monitoring of civil infrastructures using wireless sensor networks, in *Proceedings of the 6th International Conference on Information Processing in Sensor Networks* (ACM, 2007), p. 263
7. T. Nagayama, B.F. Spencer Jr, Structural health monitoring using smart sensors, *N.S.E.L. Report Series 001* (2008)
8. P.J. Wan, K.M. Alzoubi, O. Frieder, Distributed construction of connected dominating set in wireless ad hoc networks. Mob. Netw. Appl. **9**(2), 141–149 (2004)
9. O. Younis, S. Fahmy, An experimental study of routing and data aggregation in sensor networks, in *Proceedings of the IEEE International Conference on Mobile Ad-hoc and Sensor Systems* (2005), pp. 57–65
10. A.T. Zimmerman, M. Shiraishi, Automated modal parameter estimation by parallel processing within wireless monitoring systems. J. Infrastruct. Syst. **14**, 102 (2008)

Chapter 5
Design of Distributed SHM Algorithms Within WSNs-A Networked-Computing Approach

In the last chapter, a cluster-based ERA has been used to identify modal parameters. In this approach, deployed sensor nodes are divided into a number of clusters. Sampled data from each cluster are transmitted to its cluster head (CH) and each CH then implements the centralized ERA. The local results from each cluster are 'stitched' together to obtain a global one which is used to identify damage. However, one problem of this cluster-based approach is that the accuracy of damage detection obtained may not be guaranteed to be comparable with the centralized one. This is because each CH only uses its local information, which can result in the ill-conditioned problem [4], and this inaccuracy cannot be rectified via the 'stitching' process afterward.

The goal of this chapter is to design a distributed ERA algorithm which is able to achieve the same accuracy of the centralized counterpart but uses much less wireless transmission cost. We will show that designing such as a distributed algorithms serve as a guideline for more applications like SHM.

5.1 Design Objectives and the Organization of This Chapter

The desired properties the distributed ERA are summarized as

- The accuracy of the obtained results from the distributed ERA should be the same to the centralized one; and
- The required wireless transmissions and the computational load of the distributed ERA should be much smaller than the centralized one.

In this chapter, distributed ERAs are introduced which have the properties described above. For clarity, we use a step-by-step manner and one modification is described in each of the following five subsections. In Sect. 5.2, we introduce how the calculation of Markov parameters can be made distributed. Then in Sect. 5.3,

© The Author(s) 2016
J. Cao and X. Liu, *Wireless Sensor Networks for Structural Health Monitoring*,
SpringerBriefs in Electrical and Computer Engineering,
DOI 10.1007/978-3-319-29034-8_5

we describe how the most computationally expensive part in the ERA, the SVD of $\mathbf{H}(0)$, can be calculated incrementally. Based on the incremental SVD updating, we describe in Sect. 5.4 how the incremental SVD updating can be implemented under different architectures. In particular, we introduce three updating architectures, the Hamiltonian path (see Sect. 5.4.1), the minimum-connected dominating set (see Sect. 5.4.2), and the shortest path tree (see Sect. 5.4.3).

5.2 Modification 1: Calculating the Markov Parameters in a Distributed Manner

In the centralized ERA, the measured data from all the sensor nodes are first streamed to a sink, where the Markov parameters are calculated. However, since the Markov parameters are calculated based on the CCF between two signals, we can calculate the Markov parameters in a distributed way as follows. First, the reference node broadcasts its data to the whole network. After receiving the data, each sensor node then calculates the CSD and CCF to obtain its Markov parameters. Finally, the Markov parameters are transmitted back to the sink node, where they form into Hankel matrices to estimate the modal parameters. The data flow in the broadcast stage and in the aggregation stage are shown in Fig. 5.1a and b, respectively.

Compared to the centralized ERA, this modification can save significant amount of wireless transmissions. As we have described in Eq. (4.3), to obtain $N/2$ Markov parameters, each sensor need to sample at least $N/2(n_d + 1)$ data points. In the centralized ERA where the raw data are transmitted through the shortest path tree (SPT), the total data to be transmitted in a network including m sensor nodes is

$$E_{T0} = N/2(n_d + 1) \sum_{i=1}^{m} Dep^i \qquad (5.1)$$

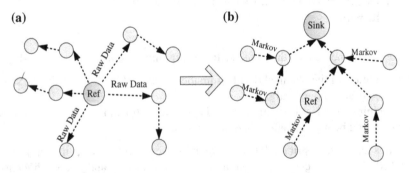

(a)   **(b)**

Fig. 5.1 Data flow of the first modification: **a** the reference node broadcasts its data, and then **b** each sensor transmits the calculated Markov parameters back to the sink

where Dep^i is the depth of the ith node in the SPT; while in the modified scheme, the required data transfer is only

$$E_{T1} = E_{bc} + N/2 \sum_{i=1}^{m} Dep^i \qquad (5.2)$$

where E_{bc} is the total data to be transmitted when broadcasting the reference data. Assume we utilize spanning tree for broadcast, then $E_{bc} = N/2(n_d + 1)(m - 1)$. Considering in our case $n_d = 20$, we have $E_{T1} \ll E_{T0}$.

However, this scheme still has some drawbacks. The sink node needs to estimate modal parameters but the associated matrix computations, particularly the SVD of the Hankel matrix $\mathbf{H}(0)$, are computationally very expensive. Consequently, a super sink node, generally a PC, is required to handle this task. Second, transmitting $N/2$ Markov parameters each time when implementing modal analysis is still a time and energy-consuming task. Therefore, we wish to find an approach which is able to calculate the modal parameters more efficiently and use less wireless transmissions.

5.3 Modification 2: Incremental SVD of the Hankel Matrix

In this section, we mainly focus on how the most expensive part in the ERA, the SVD of $\mathbf{H}(0)$, can be calculated more efficiently.

In the conventional ERA, the SVD is implemented based on the Hankel matrix constructed using all the sensor nodes. For a $\mathbf{H}(0) \in \mathbb{R}^{\alpha m \times \beta}$ matrix of with rank n, the time complexity and space complexity are $O((\alpha m)^2 \beta + \beta^2 \cdot \alpha m)$ and $O(\alpha m \cdot \beta)$, respectively [2]. To reduce the computational resource required, the SVD of $\mathbf{H}(0)$ should be implemented in an incremental manner. Incremental SVD means that, the SVD of a 'small-sized' Hankel matrix $\mathbf{H}(0)$ is calculated first which only involves data from a few sensor nodes. Then the data from remaining ones are incorporated incrementally into $\mathbf{H}(0)$ and each time the data from a new sensor is added, the SVD of the updated $\mathbf{H}(0)$ is obtained using only the previous SVD result and the newly added data.

An incremental SVD updating method is proposed in [7]. This method can solve the following problem: given the old matrix $\mathbf{A} \in \mathbb{R}^{p \times q}$ with rank n $(p, q \gg n)$, assume we already have $\{\mathbf{U}, \mathbf{S}, \mathbf{V}\}$, which are the SVD results of $\mathbf{A}$. Let $\mathbf{T} \in \mathbb{R}^{r \times q}$ be the new matrix that will be added at the bottom of $\mathbf{A}$. The incremental SVD is able to obtain $\{\mathbf{U}', \mathbf{S}', \mathbf{V}'\}$, which are the SVD results of the new matrix $\begin{pmatrix} \mathbf{A} \\ \mathbf{T} \end{pmatrix}$, using only $\{\mathbf{U}, \mathbf{S}, \mathbf{V}\}$ and $\mathbf{T}$. It should be noted that after updating, $\mathbf{S}'$ is still a diagonal matrix as $\mathbf{S}$, and both $\mathbf{U}'$ and $\mathbf{V}'$ are still unitary as were the case for $\mathbf{U}$ and $\mathbf{V}$: $\mathbf{U}'^T \mathbf{U}' = \mathbf{V}'^T \mathbf{V}' = \mathbf{I}$. According to [7], for a $\mathbf{H}(0) \in \mathbb{R}^{\alpha m \times \beta}$ matrix of with rank n, the incremental method has time complexity $O(\alpha m \cdot \beta \cdot n)$ and space complexity $O((\alpha m + \beta)n)$—much better than traditional SVD.

Unfortunately, the method cannot be directly used here to update the SVD of Hankel matrix $\mathbf{H}(0)$ since *the newly added Markov parameters from a sensor node are not appended at the bottom of the* $\mathbf{H}(0)$ *but at the bottom of each **block row** of* $\mathbf{H}(0)$ (see the definitions of Hankel matrix and Markov parameters in Eq. (4.7) and in Eq. (4.6), respectively). Modification is therefore needed.

Let us suppose that initially, we have constructed $\mathbf{H}^m(0)$, the Hankel matrix $\mathbf{H}(0)$ including m nodes:

$$
\mathbf{H}^m(0) = \begin{bmatrix} \mathbf{Y}(1) & \mathbf{Y}(\alpha+1) & \cdots & \mathbf{Y}(\beta\alpha - \alpha + 1) \\ \mathbf{Y}(2) & \mathbf{Y}(\alpha+2) & \cdots & \mathbf{Y}(\beta\alpha - \alpha + 2) \\ \vdots & & & \\ \mathbf{Y}(\alpha) & \mathbf{Y}(2\alpha) & \cdots & \mathbf{Y}(\beta\alpha) \end{bmatrix} \tag{5.3}
$$

where $\mathbf{Y}(k) = [Y^1(k), Y^2(k), \ldots, Y^m(k)]^T$ and $Y^i(k)$ is the kth Markov parameter of the ith node.

Assume the structure under test is dominated by its first n modes. The SVD of $\mathbf{H}^m(0)$ is implemented and we have $\mathbf{U}_{2n}$, $\mathbf{S}_{2n}$, and $\mathbf{V}_{2n}$ and they satisfy

$$
\mathbf{H}^m(0) = \mathbf{U}_{2n}\mathbf{S}_{2n}\mathbf{V}_{2n}^T \tag{5.4}
$$

Assume now we have Markov parameters from another sensor node $\{Y^{m+1}(1), Y^{m+1}(2), \ldots, Y^{m+1}(N/2)\}$. Then the Hankel matrix $\mathbf{H}^{m+1}(0)$ which corresponds to all these $m+1$ sensor nodes can be represented as

$$
\mathbf{H}^{m+1}(0) = \begin{bmatrix} \mathbf{Y}(1) & \mathbf{Y}(\alpha+1) & \cdot\cdot & \mathbf{Y}(\beta\alpha - \alpha + 1) \\ Y^{m+1}(1) & Y^{m+1}(\alpha+1) & \cdot\cdot & \\ \vdots & & & \\ \mathbf{Y}(\alpha) & \mathbf{Y}(2\alpha) & \cdot\cdot & \mathbf{Y}(\beta\alpha) \\ Y^{m+1}(\alpha) & Y^{m+1}(2\alpha) & \cdot\cdot & Y^{m+1}(\beta\alpha) \end{bmatrix} \tag{5.5}
$$

To use the incremental SVD updating in [7], *we need perform a series of row switching procedures on* $\mathbf{H}^{m+1}(0)$, *through which the Markov parameters of the newly added sensor node have been put at the bottom of the matrix*. These row switching procedures can be represented as a transformation matrix $\tilde{\mathbf{T}}_s \in \mathbb{R}^{\alpha(m+1)\times\alpha(m+1)}$ which satisfies:

$$
\tilde{\mathbf{T}}_s\mathbf{H}^{m+1}(0) = \begin{bmatrix} & \mathbf{H}^m(0) & \\ Y^{m+1}(1) & Y^{m+1}(\alpha+1) & \cdots\cdots \\ \vdots & & \\ Y^{m+1}(\alpha) & Y^{m+1}(2\alpha) & \cdots\cdots \end{bmatrix} \tag{5.6}
$$

Let the SVD results for $\tilde{\mathbf{T}}_s\mathbf{H}^{m+1}(0)$ be denoted as $\tilde{\mathbf{U}}_{2n}, \tilde{\mathbf{S}}_{2n}$ and $\tilde{\mathbf{V}}_{2n}$. $\{\tilde{\mathbf{U}}_{2n}, \tilde{\mathbf{S}}_{2n}, \tilde{\mathbf{V}}_{2n}\}$ satisfy

$$
\tilde{\mathbf{T}}_s\mathbf{H}^{m+1}(0) \overset{svd}{=} \tilde{\mathbf{U}}_{2n}\tilde{\mathbf{S}}_{2n}\tilde{\mathbf{V}}_{2n}^T \tag{5.7}
$$

and they can be calculated using the SVD of $\mathbf{H}^m(0)$ and the newly obtained Markov parameters $\{Y^{m+1}(1), Y^{m+1}(2), \ldots, Y^{m+1}(N/2)\}$. Multiply $\tilde{\mathbf{T}}_s^{-1}$ at both sides of Eq. (5.7), we can obtain three matrices $\mathbf{U}'_{2n}$, $\mathbf{S}'_{2n}$ and $\mathbf{U}'_{2n}$, which satisfy the following equation:

$$\mathbf{H}^{m+1}(0) = \tilde{\mathbf{T}}_s^{-1}\tilde{\mathbf{U}}_{2n}\tilde{\mathbf{S}}_{2n}\tilde{\mathbf{V}}_{2n}^T \tag{5.8}$$

Define $\mathbf{U}'_{2n} = \tilde{\mathbf{T}}_s^{-1}\tilde{\mathbf{U}}_{2n}$, $\mathbf{S}'_{2n} = \tilde{\mathbf{S}}_{2n}$, $\mathbf{V}'_{2n} = \tilde{\mathbf{V}}_{2n}$, we have

$$\mathbf{H}^{m+1}(0) = \mathbf{U}'_{2n}\mathbf{S}'_{2n}\mathbf{V}'^T_{2n} \tag{5.9}$$

Definition 3 $\mathbf{U}'_{2n}$, $\mathbf{S}'_{2n}$ and $\mathbf{V}_{2n}$ in the Eq. (5.9) are the SVD of $\mathbf{H}^{m+1}(0)$: $\mathbf{S}'_{2n}$ is a diagonal matrix, and $\mathbf{V}'^T_{2n}\mathbf{V}'_{2n} = \mathbf{U}'^T_{2n}\mathbf{U}'_{2n} = \mathbf{I}_{2n}$

Proof According to incremental SVD algorithm in [7], $\mathbf{S}'_{2n} = \tilde{\mathbf{S}}_{2n}$ is a diagonal matrix and $\mathbf{V}'_{2n} = \tilde{\mathbf{V}}_{2n}$ is composed of orthogonal columns. To prove that all the columns in $\mathbf{U}'_{2n}$ are orthogonal with each other, we first define the primary matrix switching transformation, $\mathbf{T}_{ij}$, when it multiplies with a matrix $\mathbf{A}$, switches all the matrix elements of $\mathbf{A}$ in row i with their counterparts in row j. $\mathbf{T}_{ij}$ is represented as

$$\mathbf{T}_{ij} = \begin{array}{c} \\ \\ i\text{throw} \rightarrow \\ \\ j\text{throw} \rightarrow \\ \\ \end{array} \left(\begin{array}{ccccccc} 1 & & & & & & \\ & \ddots & & & & & \\ & & 0 & \cdots & 1 & & \\ & & & \ddots & & & \\ & & 1 & \cdots & 0 & & \\ & & & & & \ddots & \\ & & & & & & 1 \end{array} \right) \tag{5.10}$$

*i*thcol   *j*thcol

By comparing Eq. 14 with Eq. 15 in the main manuscript, it can be seen that converting $\mathbf{H}^{m+1}(0)$ to $\tilde{\mathbf{T}}_s\mathbf{H}^{m+1}(0)$ can be realized using a series of simple $\mathbf{T}_{ij}$ transformations. This indicates that

$$\tilde{\mathbf{T}}_s = \prod_{k=1}^{z} \mathbf{T}_{a_k b_k} \tag{5.11}$$

According to [2], the inverse of $\mathbf{T}_{ij}$ is itself: $\mathbf{T}_{ij}^{-1} = \mathbf{T}_{ij}$, therefore, $\tilde{\mathbf{T}}_s^{-1} = \prod_{k=z}^{1} \mathbf{T}_{a_k b_k} = \tilde{\mathbf{T}}_s^T$, we have:

$$\mathbf{U}'^T_{2n}\mathbf{U}'_{2n} = (\tilde{\mathbf{T}}_s^{-1}\tilde{\mathbf{U}}_{2n})^T(\tilde{\mathbf{T}}_s^{-1}\tilde{\mathbf{U}}_{2n}) = \mathbf{I}_{2n} \tag{5.12}$$

Therefore, we have proved that $\mathbf{U}'_{2n}$, $\mathbf{S}'_{2n}$, and $\mathbf{V}'_{2n}$ are the SVD of $\mathbf{H}^{m+1}(0)$.

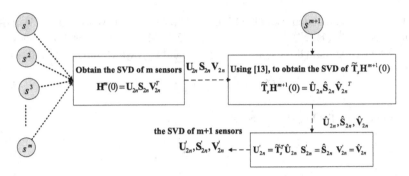

Fig. 5.2 Incrementally updating the SVD of $\mathbf{H}(0)$

In addition, we know that $\tilde{\mathbf{T}}_s^{-1} = \tilde{\mathbf{T}}_s^T$, and we can use $\mathbf{U}_{2n}' = \tilde{\mathbf{T}}_s^T \tilde{\mathbf{U}}_{2n}$ to replace $\mathbf{U}_{2n}' = \tilde{\mathbf{T}}_s^{-1} \tilde{\mathbf{U}}_{2n}$ when calculating $\mathbf{U}_{2n}'$. An even better approach can be utilized in which we do not need to find out the mathematical expression of $\tilde{\mathbf{T}}_s^{-1}$. Since $\tilde{\mathbf{T}}_s$ moves the Markov parameters of the newly added sensor node from each block row of $\mathbf{H}^{m+1}(0)$ to the last α rows, we only need to do the reverse operations on $\tilde{\mathbf{U}}_{2n}$ to obtain $\tilde{\mathbf{T}}_s^{-1} \tilde{\mathbf{U}}_{2n}$. The above procedures are summarized in Fig. 5.2.

There is one requirement about the number of sensor nodes that need to be included in the initial Hankel matrix $\mathbf{H}(0)$ on which we can start updating procedure. *We require that the number of sensor nodes included in the initial $\mathbf{H}(0)$ should be larger than 4.* The justification is as follows.

As was mentioned previously, if the vibration of a structure is dominated by its first n modes, the order of the final Hankel matrix $\mathbf{H}(0)$ is $2n$. Considering the fact that the vibration of most large civil infrastructures is dominated by only the first few modes, $n = 5$ is used in this chapter. Therefore, the rank of initial $\mathbf{H}(0)$ should be larger than 10. Since $\alpha = 20$ and $\beta = 100$ are adopted when constructing Hankel matrices, *we require that the number of sensor nodes included in the initial $\mathbf{H}(0)$ should be larger than 4.* This makes the size of the initial $\mathbf{H}(0)$ be 80-by-100, leaving enough margin to ensure the rank of initial $\mathbf{H}(0)$ meet the requirement.

5.4 Modification 3: Incremental SVD Updating in a WSN

The incremental SVD updating scheme proposed in Sect. 5.3 is still a centralized approach since it is implemented at a sink node. It is more preferable that the incremental SVD updating can be implemented within the network instead of in the sink. We will show that it is possible and by doing this, not only the computational capability of wireless sensor nodes can be fully utilized, the wireless transmissions can also be further decreased.

The key still lies in the SVD. Assume the $N/2$-length Markov parameters from each of the m nodes go through a certain node when they on the way to the sink.

Correspondingly, the total number of data to be forwarded by this node to its parent, denoted as e_{T1}, is

$$e_{T1} = \frac{mN}{2} \tag{5.13}$$

On the other hand, if this node constructs a Hankel matrix using these Markov parameters, implements the SVD, and then transmits the SVD results $\{\mathbf{U}_{2n}, \mathbf{S}_{2n}, \mathbf{V}_{2n}\}$, the data size to be transmitted, denoted as e_{T2}, would be $e_{T2} = \alpha m \cdot 2n + (2n)^2 + 2n \cdot \beta$, where the three terms correspond to the data amount of $\mathbf{U}_{2n}$, $\mathbf{S}_{2n}$, and $\mathbf{V}_{2n}$, respectively. Considering $N = 4096$, $n = 5$, $\alpha = 20$ and $\beta = 100$, it can be further approximated as

$$e_{T2} \approx \frac{mN}{20} + P, \tag{5.14}$$

where $P = (2n)^2 + 2n\beta = 1100$ is a constant. Comparing e_{T1} with e_{T2}, we can see that transmitting the SVD results is more efficient than the Markov parameters even when m is small. In the following sections, we will propose three architectures by which the SVD of $\mathbf{H}(0)$ can be updated incrementally in a WSN.

5.4.1 Updating the SVD Along a Hamiltonian Path in the Network

A straightforward way is to update the SVD along a route passing through all the sensor nodes (e.g., a Hamiltonian path if there exists one). This scheme is shown in Fig. 5.3. Initially, a number of nodes ($m \geq 3$) transmit their identified Markov parameters to a designated node. This node then constructs, from the received and its own Markov parameters, a Hankel matrix $\mathbf{H}(0)$ and implements the SVD. The SVD results $\{\mathbf{U}_{2n}, \mathbf{S}_{2n}, \mathbf{V}_{2n}\}$ are then transmitted along a path which visits each of the remaining nodes at least once. On the path, when $\{\mathbf{U}_{2n}, \mathbf{S}_{2n}, \mathbf{V}_{2n}\}$ reach one en-route node, they are updated by incorporating the its Markov parameters using the method proposed in the previous section.

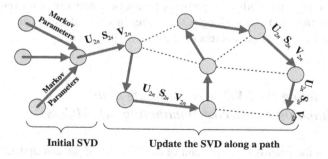

Fig. 5.3 Incrementally and distributedly updating the SVD

Ideally, if exists, the Hamiltonian path as was shown in Fig. 5.3 should be selected which visits each node exactly once. If the Hamiltonian path does not exist, we use the solution to the following problem:

Given a connected graph $G = (V, E)$, *how to find a path which includes each vertex in V at least once and the length of the path is minimized.*

We first build a minimum spanning tree (MST), then perform a depth-first tree traversal. The obtained traversal path has 2-approximation to our problem: Assume A be the optimal traversal path for visiting each node at least once, then clearly $|MST| < |A|$. While our depth-first tree traversal has length at most $2|MST| < 2|A|$.

After the $\{\mathbf{U}_{2n}, \mathbf{S}_{2n}, \mathbf{V}_{2n}\}$ reaches the last node on the path, they become the SVD of the $\mathbf{H}(0)$ of all the sensor nodes in the network. However, to obtain the modal parameters, we still need another Hankel matrix $\mathbf{H}(1)$. To find $\mathbf{H}(1)$, we first reconstruct $\mathbf{H}(0)$ using the final updated SVD by Eq. (4.9). By observing Eq. (4.7), we can found that $\mathbf{H}(0)$ already contains all the Markov parameters needed in $\mathbf{H}(1)$ except the last term: $\mathbf{Y}(\beta\alpha+1)$. Therefore, each sensor node in the network only need to transmit its $Y^i(\beta\alpha + 1)$ to the last node and the latter identifies modal parameters of the structure accordingly. Considering the number of transmission associated with $\mathbf{Y}(\beta\alpha + 1)$ is small, they are omitted in the remaining of the chapter.

We calculate the overall amount of data that need to be wirelessly transmitted in the network using this approach. In an ideal condition when a Hamiltonian path exists for a network with a total of m sensor nodes, the number of transmissions, denoted as E_{T2} is

$$E_{T2} = 3 \cdot N/2 + \sum_{k=4}^{m-1} \left(\frac{N}{20}k + P \right) \tag{5.15}$$

where the first and second terms at the right side of Eq. (5.15) correspond to the number of transmissions when the initial Hankel matrix is constructed and when $\{\mathbf{U}_{2n}, \mathbf{S}_{2n}, \mathbf{V}_{2n}\}$ travel along the Hamiltonian path, respectively. According to Eq. (5.15), the overall wireless transmissions quadratically increase with the number of sensor nodes. This is because once $\{\mathbf{U}_{2n}, \mathbf{S}_{2n}, \mathbf{V}_{2n}\}$ meet a node on the path and are updated by incorporating its Markov parameters, the size of $\{\mathbf{U}_{2n}, \mathbf{S}_{2n}, \mathbf{V}_{2n}\}$ will be increased by $\frac{N}{20}$. Therefore, the amount of data to be transmitted along a path is expected to be small initially but can be significantly large if the length of the path is large enough. In some conditions, particularly in a large network, it is even possible that updating the SVD along a long path requires more wireless transmissions than transmitting the Markov parameters through the SPT.

5.4.2   Updating the SVD Along a Route in the Minimum-Connected Dominating Set (MCDS)

According to the discussion at the end of the Sect. 5.4, if we can update the SVD along a route shorter than the Hamiltonian path, the required wireless transmissions

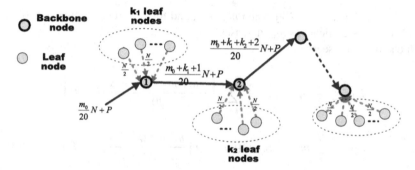

Fig. 5.4 Update the SVD along backbone nodes while the leaf nodes transmit the Markov parameters to the corresponding backbones

may be decreased. Assume a sensor node s^0 has received the updated $\{\mathbf{U}_{2n}, \mathbf{S}_{2n}, \mathbf{V}_{2n}\}$ including all the m_0 en-route nodes so far. Also assume there are still m_r sensor nodes, including s^0, are to be included. If the SVD is updated through the Hamiltonian path among these m_r sensor nodes, the total number of data to be transmitted, starting from s^0, will be

$$E_{T2} = \sum_{k=1}^{m_r-1} \left[\frac{(m_0 + k)N}{20} + P \right] \tag{5.16}$$

On the other hand, we construct *a path which satisfies the condition that all the* m_r *sensor nodes are either on the path or have neighbors on the path.* For convenience, we call the nodes on the path the **'backbone'** nodes and the rest the **'leaf'** nodes. After the path is established, the leaf nodes transmit their data to their corresponding backbone neighbors. As before, $\{\mathbf{U}_{2n}, \mathbf{S}_{2n}, \mathbf{V}_{2n}\}$ are updated when they travel along the path but when they meet one backbone node, they will incorporate the Markov parameters of both this node and all of its leaf nodes. The above process is shown in Fig. 5.4.

Still let the total number of nodes whose data are to be included be m_r, and among which, we identify m_b backbone nodes and $m_r - m_b$ leaf nodes. The amount of data transmitted in this scheme is

$$E_{T3} = N/2(m_r - m_b) + \sum_{i=1}^{m_b-1} \left(\frac{\left(m_0 + \sum_{j=1}^{i} k_j \right) N}{20} + P \right) \tag{5.17}$$

where $k_i (i = 1, \ldots, m_b)$ is number of leaf nodes for ith backbone node. Obviously, $\sum_{i=1}^{m_b} k_i = m_r - m_b$. The first term corresponds to the number of data transmitted from the leaf nodes to the backbones, and the second, third, and the last term correspond to the number of data transmitted from the 1st, 2nd and the next-to-last backbone nodes, respectively. Note that the backbone node at the end of the path does not need to further transmit the SVD results.

We prove that $E_{T2} > E_{T3}$ when $m_0 > 5$, and the difference between E_{T2} and E_{T3} becomes significant as m_0 increases.

It can be easily seen

$$E_{T3} \leq \frac{N}{2}(m_r - m_b) + \sum_{k=1}^{m_b-1} \left[\frac{m_0 + m_r - m_b + k}{20} N + P \right] \qquad (5.18)$$

$$E_{T3} \geq \frac{N}{2}(m_r - m_b) + (m_b - 1) \left[\frac{m_0 + m_r - m_b}{20} N + P \right] \qquad (5.19)$$

The maximum and minimum values in of E_{T3} shown in Eqs. (5.18) and (5.19) can be achieved when $\{k_1 = m_r - m_b, k_2 = k_3 = \cdots = 0\}$, and when $\{k_1 = k_2 = k_{m_b-1} = 0, k_{m_b} = m_r - m_b\}$, respectively.

Moreover, comparing the total data to be transmitted in these schemes, we have

$$E_{T2} - E_{T3} = (m_r - m_b) \left(\frac{m_0 N}{20} + P - \frac{N}{2} \right) + \frac{N}{20} \sum_{k=m_b}^{m_r-1} k$$
$$- \frac{N}{20} \left[(k_1) + (k_1 + k_2) + \cdots + (k_1 + \ldots k_{m_b-1}) \right] \qquad (5.20)$$

Using Eq. (5.18), it can be proved that

$$E_{T2} - E_{T3} \geq (m_r - m_b) \left(\frac{m_0 N}{20} + P - \frac{N}{2} \right) + \frac{N}{40}(m_r - m_b)(m_r - m_b + 1)$$

Therefore, we have $E_{T2} > E_{T3}$ when $m_0 > 5$. Also can be seen is that the difference between E_{T2} and E_{T3} becomes significant as m_0 increases.

Inspired by the observation above, we can utilize the minimum connected dominating set (MCDS) to find out the path along which the USV results are updated. This is illustrated in Fig. 5.5. Initially, a sensor s^0 has the SVD of a Hankel matrix

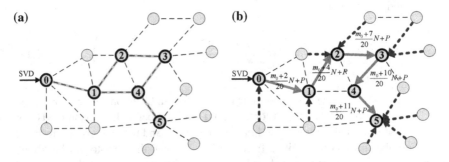

Fig. 5.5 Example of using MCDS to update the SVD. **a** The original topology and the graph of MCDS. **b** SVD update along a Hamiltonian path in the MCDS

including a few sensor nodes. Based on the network topology of all the remaining nodes, including s^0, the MCDS is found. Then a path which includes all the nodes in the MCDS is found. If exists, the Hamiltonian path should be chosen. Otherwise, we construct a MST of the MCDS, and then perform a depth-first tree traversal to find the path.

It should be noted that if a leaf node has more than one backbone neighbors in the MCDS, it should choose the one nearest to path end to forward its data. This is because the size of the SVD accumulates every time data from a new sensor is incorporated, and delay of the incorporating data from a new sensor is always preferable in terms of the wireless transmissions.

5.4.3   Modification 5: Updating the SVD Along the Shortest Path Tree

In the previous sections, the SVD is updated along a path (e.g., a Hamiltonian path either in the original graph or in the MCDS). Using a path has one potential problem particularly in a sparse network since a Hamiltonian path in the original graph or in the MCDS may not exist. This can result in a longer path and the associated high communication cost. However, if we can find another function which allows a set of SVD results to be merged together, we can adopt the **tree** architecture widely used in data aggregation in WSNs. In this section, we propose a method to merge a set of SVD results and based on which describe how the SVD results can be updated along the SPT. We will show later that this approach can achieve a constant approximation ratio. In addition, compared with the approaches in which the SVD are updated along a path, many computations when updating along a tree can be implemented in parallel, which significantly decreases the time required for the ERA.

To merge a set of SVD results is very simple. Assume one sensor node receives two sets of SVDs, $\{U_{2n}, S_{2n}, V_{2n}\}$ and $\{\tilde{U}_{2n}, \tilde{S}_{2n}, \tilde{V}_{2n}\}$, each corresponding to a cluster of sensor nodes c_1 and c_2, respectively. To obtain the SVD for $c_1 \bigcup c_2$, one of the SVD set, for example $\{\tilde{U}_{2n}, \tilde{S}_{2n}, \tilde{V}_{2n}\}$, is converted back to the Hankel matrix using Eq. (5.21).

$$\mathbf{H}^{c2}(0) = \tilde{\mathbf{U}}_{2n}\tilde{\mathbf{S}}_{2n}\tilde{\mathbf{V}}_{2n}^T \tag{5.21}$$

where $\mathbf{H}^{c2}(0)$ is the Hankel matrix corresponding to cluster c_2. According to the definition of Hankel matrix shown in Eq. (5.3), the Markov parameters of each node in c_2 can be simply extracted from $\mathbf{H}^{c2}(0)$. Then what we have is $\{U_{2n}, S_{2n}, V_{2n}\}$ of cluster c_1, and the Markov parameters of cluster c_2. $\{U_{2n}, S_{2n}, V_{2n}\}$ can be incrementally updated by incorporating c_2's data. Note that when we decide which SVD set should be updated, the one with larger size should be chosen since this would result in lower computation.

In this scheme, based on the network topology, the SPT is established with the root being the graph's topology center. Then the SVD is implemented in a bottom-up manner. For a leaf node at the bottom of the tree, it transmits its Markov parameters to its parent. For a non-leaf node, the types of received data can be (1) Markov parameters only, (2) Markov parameters with single SVD results, and (3) with multiple SVD results. In case of condition 1, the node either constructs a Hankel matrix and implements the SVD (if the number of its children is equal or larger than 3), or simply packs the received Markov parameters with its own data and delivers them to its parent. In case of condition 2, the SVD results are updated using its own and the received Markov parameters. In case of condition 3, the SVD results are merged first and then updated using the Markov parameters. The results, either in the form of Markov parameters or in the SVD results, are transmitted to its parent. Eventually, this procedure ends at the root. The actions that a non-leaf node takes when it receives data from its children are described in Algorithm 5.

Algorithm 5 The actions that a non-leaf node takes when it receives data from its children

---

1: **if** *it receives Markov para.* **then**
2:     **if** the number of children nodes $\geq$ 3 **then**
3:         it constructs a Hankel matrix including all the received Markov para. and its own data and then implements **F({Markov parameters}) = SVD results**.
4:     **else**
5:         it packs its own Markov para. with the received data.
6:     **end if**
7: **else**
8:     **if** *it receives SVDs or SVDs + Markov para.* **then**
9:         Using its own and the received Markov para., it updates the SVD by implementing **F(Markov para., SVD results) = SVD results**.
10:     **else**
11:         **if** *it receives multiple SVD results* **then**
12:             it merges these SVDs first by **F({SVD results}) = SVD results** and then updates the SVD using its own Markov parameters by **F(Markov para., SVD results) = SVD results**.
13:         **end if**
14:     **end if**
15: **end if**

To illustrate this updating procedure, we use a simple example in Fig. 5.6 where the SPT is established. Initially, nodes in the bottom layer (we call it the 1st layer) of the tree transmit their Markov parameters to their parents. For nodes at the 2nd layer (i.e., Nodes 8, 6, 7, 3, 4), 8 and 4 do not receive enough Markov parameters to initiate the SVD, and therefore have to transmit the received data along with its own Markov parameters to their parents. Nodes 3, 6 and 7 are able to implement the SVD and transmit the SVD results to higher level nodes. For nodes in the higher level, node 2 receives two SVD sets from nodes 6 and 7, respectively, and it will merge them and update it using its own Markov parameters.

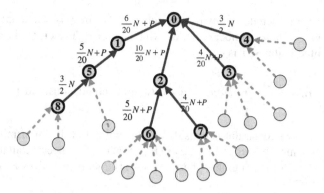

Fig. 5.6 Using a SPT to update the SVD

For updating along the SPT, we have the following theorem:

Definition 4 The approximation ratio of using the SPT to update the SVD is 10.

Proof Assume there exists an optimal updating scheme by which all the Markov parameters of all the sensors are aggregated to the sink node to obtain the SVD which incorporating information from all the sensor nodes. The amount of data to be transmitted in this scheme is denoted as E_{opt}. We first prove that

$$E_{opt} > \frac{N}{20} \sum_{i=1}^{m} Dep^i + m * P \qquad (5.22)$$

where Dep^i is the depth of node i in the SPT and m is the total number of nodes.

We first remove the initial constraint, which requires that the initial SVD can only be calculated when we have data from at least four sensors, and assume any sensor can implement the SVD. Therefore, for a leaf node, the amount of data it transmits to its parent is $N/20 + P$. For a non-leaf node which receives the SVD result containing m sensors, the amount of updated SVD result transmitted from this node is $(m + 1) * N/20 + P$. Therefore, if we let $P = 0$, then the original network can be regarded as a new network in which each sensor generates $N/20$ data, and data from all the sensors need to be delivered to the sink without any aggregation along the route. The optimal routing architecture must be the SPT. The total amount of data transmitted in this SPT is

$$E_{opt} = \frac{N}{20} \sum_{i=1}^{m} Dep^i \text{ (assume no ini. constraint and } P = 0) \qquad (5.23)$$

When P is considered, for any routing schemes, the increased amount of data is only associated with the number of nodes, which is $m * P$. Therefore, the amount of data transmitted in the SPT is

$$E_{opt} = \frac{N}{20} \sum_{i=1}^{m} Dep^i + m * P \text{ (assume no ini. constraint)} \qquad (5.24)$$

When the initial constraint is considered, the amount of data transmitted in the optimal scheme must be larger than E_{opt} in Eq. (5.24). This is because all the sensors must transmit at least $N/2$ number of data, which is larger than $N/20 + P$. Therefore, we have Eq. (5.22).

We will prove that using the SPT and the Algorithm 5, the amount of data to be transmitted in the network, denoted as E_{T4}, satisfies

$$E_{T4} \leq \frac{N}{2} \left(\sum_{i=1}^{m} Dep^i \right), \qquad (5.25)$$

To prove Eq. (5.25), we classify nodes in the network into two categories according to the number of children they have. For convenience, nodes in the first category have no more than two children and are called as A-nodes. The remaining nodes are called B-nodes. Let the number of A-nodes be p, then we have $m - p$ B-nodes. It can be easily seen that the total amount of data transmitted by the A-nodes is $E_A = \frac{N}{2} \left(\sum_{i \in A} \overline{Dep}^i \right)$, where $\overline{Dep}^i$ is the hop count between node i and its nearest B-node.

On the other hand, the amount of wireless transmissions of the B-nodes can be expressed as $E_B = \frac{N}{20} \left(\sum_{i=1}^{m} Dep^i \right) - \frac{N}{20} \left(\sum_{i \in A} \overline{Dep}^i \right) + (m - p) * P$. Therefore, the total amount of transmissions is

$$E_{T4} = E_A + E_B = \frac{N}{20} \left(\sum_{i=1}^{m} Dep^i \right) + \frac{9N}{20} \left(\sum_{i \in A} \overline{Dep}^i \right) + (m - p) * P$$

E_{T4} can reach the maximum when the sink becomes the only B-node. In this condition: $\overline{Dep}^i = Dep^i$, $E_3 = \frac{N}{2} \left(\sum_{i=1}^{m} Dep^i \right)$. Considering

$$E_{T4} \leq \frac{N}{2} \left(\sum_{i=1}^{m} Dep^i \right), E_{opt} > \frac{N}{20} \sum_{i=1}^{m} Dep^i$$

We have

$$E_{T4} \leq 10 E_{opt}$$

5.5 Summary of Communication and Computation Cost of the Proposed Methods

In this section, we compare the communication and computation cost of the methods proposed so far. For convenience, we first define the schemes described in the previous sections.

1. **Scheme 1**: All the raw data are transmitted to the sink through the SPT (i.e., the centralized ERA).
2. **Scheme 2**: The reference node broadcasts its data to the network, and then all the sensors transmit the calculated Markov parameters back to the sink through the SPT (as described in Sect. 5.2).
3. **Scheme 3**: The SVD is updated along a path (described in Sect. 5.4.1).
4. **Scheme 4**: The SVD is updated along a path in the MCDS (as described in Sect. 5.4.2).
5. **Scheme 5**: The SVD is updated along the SPT (as described in Sect. 5.4.3).

Table 5.1 summarizes the communication and computation cost of the five schemes. Note that for the Scheme 2–5, the communication overhead when the sink node broadcasts its data to all other nodes is also included. We can see that compared to the conventional centralized approach (Scheme 1), the communication cost for our proposed schemes is much lower. The communication cost for Scheme 2–5 depends on different network topology. In addition, according to [7], the computation complexity for the conventional centralized is $O((\alpha m)^2 \beta + \beta^2 \alpha m)$, which is $O(m^2)$. On the other hand, since Scheme 2–5 utilize the incremental SVD updating method proposed in Sect. 5.3, the computation complexity of these schemes is only $O(m)$.

Table 5.1 Comparing the communication and computation cost for different approaches

Scheme #	Communication cost	Computaion complexity
1	$N/2(n_d + 1) \sum_{i=1}^{m} Dep^i$	$O(m^2)$
2	$N/2(n_d + 1)(m - 1) + N/2 \sum_{i=1}^{m} Dep^i$	$O(m)$
3	$N/2(n_d + 1)(m - 1) + 3 \cdot N/2 + \sum_{k=4}^{m-1} \left(\frac{N}{20}k + P\right)$	$O(m)$
4	$N/2(n_d + 1)(m - 1) + 3 \cdot N/2 + N/2(m - 3 - m_b)$ $+ \sum_{i=1}^{m_b - 1} \left(\frac{\left(3 + \sum_{j=1}^{i} k_j\right)N}{20} + P\right)$	$O(m)$
5	$N/2(n_d + 1)(m - 1) +$ $\frac{N}{20} \sum_{i=1}^{m} Dep^i + \frac{9N}{20} \sum_{i \in A} \overline{Dep}^i + (m - p) * P$	$O(m)$

5.6 Performance Evaluation

We first use simulation data to demonstrate the advantage of distributed ERA in terms of wireless communications. We consider implementing the ERA using five different schemes defined in Sect. 5.5.

Two simulation scenarios are created to evaluate the performance of these schemes in different network density and different network size, respectively. Assume we have a total of m sensor nodes which are randomly deployed on an area and each node has the same communication range R. We first fix the number of nodes $m = 100$ but gradually increase the network density by increasing R from 12 to 40 m. The area in this scenario is fixed to be 100×100. Then we maintain the network density at a certain level but gradually increase the network size by increasing m from 100 to 800. Note that to maintain the network density, when increasing m, the size of the deployment area also needs to be increased. In each scenario, a total of 500 simulations are performed and the average energy consumption of each scheme is calculated.

The results of the first scenario (i.e., in different network densities) are illustrated in Fig. 5.7. Note that for clear comparison, the logarithmic scaling for the vertical axis is utilized. It can be seen that for all these five schemes, the amount of data transmitted for the ERA is decreasing with the increase of the communication range. In addition, using Scheme 2–5 saves significant wireless transmissions compared with Scheme 1. Furthermore, when the transmission range is large enough ($>15m$ in this case), the amount of data transfer required in Scheme 3 > Scheme 2 > Scheme 4 > Scheme 5.

Likewise, Fig. 5.8 compare the data transmitted in different schemes under various network size. In terms of the data amount to be transmitted, Scheme 1 > Scheme 3 > Scheme 2 > Scheme 4 > Scheme 5.

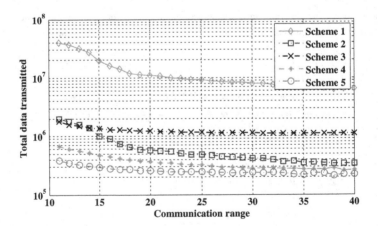

Fig. 5.7 The total data transmitted under various network densities

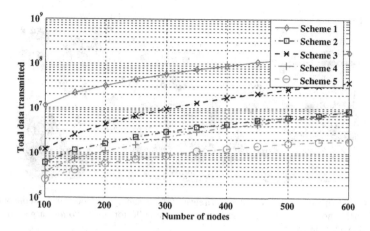

Fig. 5.8 The total data transmitted under various network size

An important property that can be seen from these figures is that compared to the Scheme 1 in which raw data are transmitted, the advantage of the distributed schemes, particularly the Scheme 5, becomes more significant in a sparse network (i.e., small R) with large number of sensor nodes (i.e., large m). This property is very favorable for SHM since this matches the real condition when wireless sensor nodes are deployed to monitor the condition of large civil infrastructures.

Using a real experiment, we have tested the performance of the best three schemes described in Sect. 5.5 (i.e., Scheme 2, Scheme 4, and Scheme 5). We deployed a number of SHM motes in the LSK Building to measure its vibration under ambient environment (see Fig. 5.9a). SHM motes are particularly developed by us for SHM. SHM motes contain relatively powerful floating point DSP processor TMS320F2812

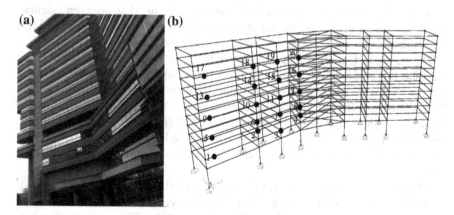

Fig. 5.9 The LSK building and measurement locations. **a** The LSK building. **b** 20 measurement locations

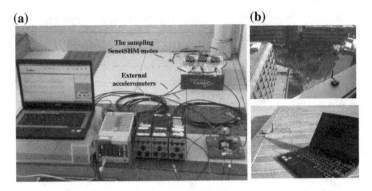

Fig. 5.10 Experiment setup. **a** Wired and wireless systems deployed at location 17 in the building. **b** *Top* a SHM mote deployed near the window as the collector node; *Bottom* a gateway mote connected to a laptop computer

running at 150 MHz and hence able to implement complicated SHM computations in much faster than most off-the-shelf wireless sensor nodes.

The numbering of the measured locations is shown in Fig. 5.9b. The system setup on measurement location 17 is illustrated in Fig. 5.10a. At each measurement location, we use three high-sensitive external sensors named KD1300 to record the vibration (see Fig. 5.10a) in three directions. Signal recorded at KD1300 is amplified and then fed into a SHM mote where they are stored as 8-byte single-precision floating point format. For convenience, we call the SHM motes connecting to the KD1300 the sampling motes. In this experiment, wireless communication is not directly established among the sampling motes at different locations since they belong to different rooms and are not able to directly communicate well. To solve this problem, we deploy a particular SHM mote acting as local data collector near the window of each location (see the top figure of Fig. 5.10b). Vibrational data sampled at the sampling motes are transmitted to this collector first. These 20 collector nodes can be regarded as independent wireless sensor nodes in this WSN. In this chapter, when calculating the communication cost, we assume the collector nodes already have the local vibration data and only consider the communication among the collector nodes.

During the test, we first find the network topology of the collectors by running the collecting tree protocol (CTP) [3]. The topology information is transmitted to a gateway node connected to a laptop computer (see the bottom figure of Fig. 5.10b). The network topology of these 20 collector nodes is illustrated in Fig. 5.11a. From the topology, the laptop computer calculates the SPT (for Scheme 2 and Scheme 5), or the path in the MCDS (for Scheme 4). For each node in the network, its children and parent are then determined. This information is then broadcast to all the collector nodes.

All the deployed sampling motes are then synchronized using the modified flooding time synchronization protocol (FTSP) [6]. At a certain global time point, all the sampling motes start sensing with sampling rate of 1024 Hz. The sampling procedure lasts for 50 sec.

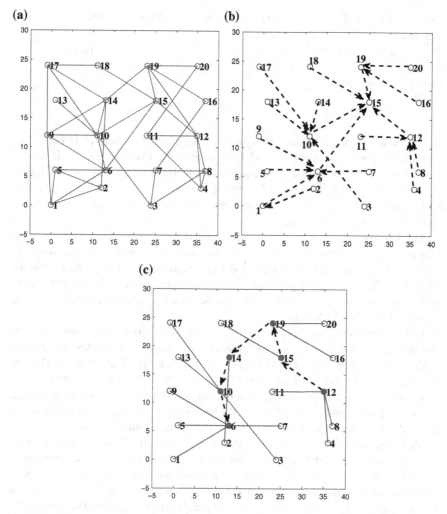

Fig. 5.11 a The network topology of collector nodes **b** The SPT (with root 14) and **c** the path in the MCDS (shown as *red dots* in the figure)

After all the collector nodes have obtained the sampled data from their corresponding sampling motes, the network center No. 15, is taken as the reference node to broadcast the raw data to the network. Each of the remaining nodes, once receives the data, calculates its Markov parameters.

Both Scheme 2 and Scheme 5 adopt SPT rooted at node 15, and therefore they share the same routing shown in Fig. 5.11b. The differences in these two schemes is that in the former, the Markov parameters are transmitted while in the latter, the SVD

Table 5.2 Theoretical data amount and the actual time cost in different schemes

Schemes #	1	2	4	5
Theoretical data amount	4515840	215040	119164	103804
Actual time cost (min)	N/A	38.2	26.1	24

is updated along the SPT. The routing used by Scheme 4 is illustrated in Fig. 5.11c. Note that arrows with black dashed lines connect backbone nodes while arrows with blue dashed-dot lines connect leaf nodes to backbone nodes. In this figure, a Hamiltonian path is found which connects all the backbone nodes together. The SVD results will finally reach node 6, where the modal parameters are calculated and transmitted to the root node 15.

The theoretical amount of data transmitted in these three schemes, along with the Scheme 1 in which all the raw data are transmitted directly to the sink, are shown in Table 5.2. Updating the SVD along the SPT (the Scheme 5) and along the path in the MCDS (the Scheme 4) are significantly better than delivering the Markov parameters to the sink (the Scheme 2). In addition, the Scheme 5 requires less amount of data transmissions than the Scheme 4.

In this experiment, we also compare the actual time it takes for different schemes. For Scheme 2, the time it took to finish the transmission of Markov parameters is about 38.2 min. While the time it takes for updating the SVD in the MCDS (the Scheme 4) and along the SPT (the Scheme 5) are about 26.1 and 24 min, respectively. It should be noted that the computation time in these two schemes is also included. Considering when using the Scheme 1, the total amount of data is about 20 times of Scheme 2, the advantage of the distributed ERA is thus quite obvious.

More importantly, we will show that the distributed ERAs are able to give extremely accurate modal parameters as was in the traditional centralized method. Natural frequencies, and mode shapes of the Scheme 5 are compared with those calculated on the laptop computer connected with the gateway using the received Markov parameters. The comparison results are shown in Table 5.3. The error in the frequency is calculated as the difference between the estimates on the SHM Mote and the PC. The estimation error in the mode shapes is investigated in terms of the modal assurance criterion (MAC) [1]. As shown in Table 5.3, the modal parameters in the SHM Mote and those on the PC are almost identical. Also, for comparison purpose, the modal parameters identified from the clustering-based ERA proposed in [5] also shown in Table 5.3. Our method can achieve much higher accuracy than the cluster-based method.

Table 5.3 Comparing accuracy of our approach to the cluster-based ERA

Mode	Our distributed ERA			Cluster-based ERA		
	Natural frequency		Mode shapes	Natural frequency		Mode shapes
	f (Hz)	Diff. (10^{-4})	1-MAC	f (Hz) (10^{-6})	Diff. (10^{-2})	1-MAC (10^{-2})
1	0.7536	0.3241	1.2844	0.7512	0.32	2.623
2	0.9544	1.2712	1.7453	1.0130	3.09	0.3432
3	1.3246	1.3811	2.9254	1.3381	1.45	2.289
4	6.7754	1.9245	3.5429	6.9223	2.88	1.230
5	15.8815	2.6818	4.1643	16.10	1.31	3.343

It can be seen that the identified modal parameters of our approach have less than 0.03 % identification error, while using the cluster-based ERA can result in up to 3 % identification error

5.6.1 Conclusion

In this chapter, for a chosen SHM algorithm, the ERA, we proposed a few distributed versions that can be implemented in WSNs. Through simulation and experiment, the effectiveness and efficiency of the distributed schemes in terms of required wireless transmissions, computation complexity, and the accuracy of the output are demonstrated. In addition, we believe the proposed schemes can serve as a guideline for more applications of WSNs which are data intensive and involve sophisticated signal processing like SHM.

References

1. R.J. Allemang, D.L. Brown, A correlation coefficient for modal vector analysis, in Proceedings of the 1st International Modal Analysis Conference, vol. 1 (1982), pp. 110–116
2. G.H. Golub, C.F. Van Loan, *Matrix computations* (Johns Hopkins studies in mathematical sciences) (Johns Hopkins University Press, Baltimore, 1996)
3. K. Jamieson, S. Kim, P. Levis, R. Fonseca, O. Gnawali, A.Woo, TEP 123: Collection Tree Protocol, http://www.tinyos.net/tinyos-2.x/doc/
4. J.N. Juang, R.S. Pappa, Eigensystem realization algorithm for modal parameter identification and model reduction. J. Guid. Control Dyn. **8**(5), 620–627 (1985)
5. X. Liu, J. Cao et al., Energy efficient clustering for wsn-based structural health monitoring. IEEE INFOCOM **2**, 1028–1037 (2011)
6. M. Maróti, B. Kusy, G. Simon, Á. Lédeczi, The flooding time synchronization protocol, in Proceedings of the 2nd International Conference on ENSS (2004), pp. 39–49
7. H. Zha, H.D. Simon, On updating problems in latent semantic indexing. SIAM J. Sci. Comput. **21**, 782 (1999)

Chapter 6
Realizing Fault-Tolerant SHM in WSNs

6.1 Introduction

One important application of wireless sensor networks (WSNs) is event detection: nodes are tasked to determine whether a particular event of interest is occurring in their sensing range. However, when a large number of low-cost wireless sensor nodes are deployed under harsh environment for a long period of time, many of them are likely to exhibit different types of faults. It is important to guarantee that the event of interest is still able to be reliably detected in the presence of faulty nodes.

In general, faults in wireless sensor nodes can be largely divided as 'function fault' and 'data faulty'. Function fault generally includes the crash of individual nodes. While node with data fault behaves normally in all aspects except for its readings. Compared with function fault, data fault is much more difficult to be detected and is likely to produce false positives. *In this chapter, without specified otherwise, 'faulty nodes' are denoted as nodes with data fault.*

Fault-tolerant event detection has been studied extensively in WSNs and many schemes have been proposed. In these schemes, the effect of faulty readings are eliminated or 'masked' through high-level fusion techniques such as majority voting [2, 8], Bayesian networks [11] etc. However, the events of interest in these schemes are relatively simple and to apply high-level fusion, it is assumed that each sensor node, if healthy, is able to give correct or at least statistically correct local answer to the occurrence of event. Because of these assumptions, they cannot be applied to SHM application.

In every collection period of SHM, sensor nodes collect the responses of structure and transmit them to a server where a SHM algorithm is implemented to detect possible event in SHM (i.e., structural damage). Different from other applications like environmental monitoring where each node can give an answer to the occurrence of event, detecting event in SHM requires low-level collaboration from multiple sensors [4]. Low-level collaboration means that raw data from multiple sensors are

© The Author(s) 2016
J. Cao and X. Liu, *Wireless Sensor Networks for Structural Health Monitoring*,
SpringerBriefs in Electrical and Computer Engineering,
DOI 10.1007/978-3-319-29034-8_6

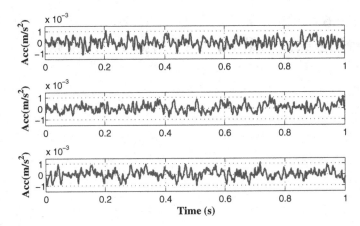

Fig. 6.1 The time history data from three sensor nodes

processed simultaneously, generally through various matrix computations such as
system identification, singular value decomposition, etc. This indicates that once
faulty data are involved in event detection, their effect cannot be eliminated any
more.

To identify nodes with faulty readings, existing approaches generally rely on comparison. A node is considered as faulty if its local decision about event is different
from the majority, its measurement is statistically different from the neighborhood
or does not match what it is supposed to be. However, most existing approaches only
take Boolean or scalar sensor data as input. While in SHM, the data sampled from
each sensor is a sequence of vibrational data. Figure 6.1 shows the data measured
from three sensors deployed on a structure. It can be seen that directly comparing
these dynamic sequences is not applicable. Extracting and comparing features contained in the measured sequences is not an easy task since features in these dynamic
sequence is no longer maximum, mean, medium, etc., but some domain associated
characteristics, and are generally in the form of feature vectors. And we will show
that these feature vectors can have '*element mismatch problem* (EMP)', where comparable elements in these vectors are located at unknown different positions.

In this chapter, we proposed FTED, a fault-tolerant event-detection scheme for
SHM. In this scheme, a method called I-FUND is designed to detect nodes with faulty
readings. I-FUND can take vector input, and can also handle the EMP. I-FUND can
be easily generalized to other applications other than SHM which share the similar
properties. In FTED, both fault node detection and event detection are distributed
and based on optimized clusters. The contribution of our work in this chapter is
summarized as follows:

- We propose a fault-tolerant and a distributed approach called FTED to detect complex event in SHM.
- We designed I-FUND to detect nodes with faulty readings. I-FUND can take vector input and handle the EMP.
- The effectiveness of FTED and I-FUND have been demonstrated through both simulation and experiment.

6.2 Related Works

Existing approaches to detect sensor nodes with faulty data can be largely divided into two categories. Approaches in the first category determine whether a sensor node is faulty by its answer to a specific question: whether the event occurs. If a node, according to its data only, makes an opposite conclusion with the majority, then it is labeled as faulty. Since approaches in this category detect a faulty node by its '0/1' answer to the event occurrence, they are called '*0/1 event-based methods*' (**0/1-Eve**). An example of 0/1-Eve is decision fusion [2]. If event or target occurs or appears, each sensor observes it via measured signal and makes a local decision which is a binary value. The local decisions are sent to a fusion center, where a global decision is made by combining these local decisions through majority voting or other fusion techniques. Then, according to whether the local decision is the same with this global, faulty nodes can be determined. To make the above strategy more applicable for WSNs, a distributed version of decision fusion was proposed in [8]. In this algorithm, each sensor node only communicates with its neighbors and collects their binary decisions. If most neighbour nodes have the same value as its particular value, a node considers that its sensor is correct. This scheme is further improved in [11] by using only part of neighbour nodes in fusion so as to further decrease the energy consumption and traffic congestion.

Despite of its simplicity, using 0/1-Eve such as decision fusion has some limitations. First, to give a local answer about whether the event occurs, each node needs to compare current sensor readings with a threshold, which is hard to be correctly determined. Also, this local decision is the preprocessing results of the actual measured data and detection over binary local decision represents the second round approximation [3].

To solve the problem, approaches in the second category do not rely on detecting the event but by comparing the measurement from sensor nodes directly. A node is identified as faulty if its measurement is statistically different from others, particularly the neighborhood, or does not match what it is supposed to be. Since in this category, measurement from each node is a scalar, these methods can be called as '*scalar measurement-based methods*' (**x-Mea**). For example, Ding et al. [3] proposes a distributed method to detect faulty nodes by determining if the difference between a node's reading and its neighbors' is significant. This is based on the premise that data collected from nodes, particularly nodes in the neighborhood, follow the same distribution. Guo et al. [6] observed that this assumption may not hold since many sensing readings (e.g., acoustic volume and thermal radiation) in sensor networks attenuate over distance. Accordingly, they proposed a detection approach for discovering data faults called FIND. FIND only assumes that the readings of sensor nodes can generally reflect the relative distance from the nodes to the event. If the readings from a node violate this rule, it is regarded as faulty.

Although the approaches described above have been successfully applied for many applications of WSNs, they still have some limitations. Both 0/1-Eve and x-Mea can only take scalar input of sensor data but in some applications of WSNs such as SHM, each node can generate a sequence of data in a single monitoring period. How to find out whether a sequence of data is faulty in WSNs has not been studied in literature before. In addition, some intrinsic assumptions in 0/1-Eve and x-Mea such as 'a global correct answer can still be obtained when data from faulty nodes are involved', 'each healthy node is able to give correct answer to the occurrence of event' are not valid for applications such as SHM where detecting event of interest requires data-level collaboration from multiple sensor nodes. Furthermore, particularly for the x-Mea, after faulty nodes have been detected, the main task of WSNs, to monitor event of interest, must be implemented from the scratch using data from remaining healthy ones. The cost paid on detecting faulty nodes generally cannot be neglected compared with detecting event itself.

Interestingly, there is also some previous work in civil engineering community to detect sensors which give faulty readings. In [5], two approaches were proposed which based on the comparison between the subspace of response and the subspace generated by the lower modes of a structural model. In [7, 14], the detection of sensor failures relies on principle component analysis (PCA). However, these approaches are not able to discriminate between sensor fault and structural damage. In addition, they are all centralized, computational intensive, and therefore, are not suitable for a resource-limited wireless sensor nodes.

6.3 FTED: The Fault-Tolerant Event Detection Scheme for WSN-Based SHM

The proposed FTED can be largely divided into three stages: (1) distributed extraction of features for faulty node detection, (2) iterative faulty node detection and (3) event detection. Two types of vibrational features, namely natural frequency and mode shape, are used in (1) and (3), respectively. Table 6.1 shows the notations used in this chapter.

Table 6.1 Summary of notations

$\mathbf{f}, \boldsymbol{\Psi}$	Natural frequency set and mode shape matrix
δm	The ratio of number of faulty and healthy sensors
v, w	The random noise on the healthy/faulty nodes, resp.
$\mu_v, \sigma_v, \mu_w, \sigma_w$	The mean and std of v and w, resp.
$\mathbf{f}_{true}, \mathbf{f}_H, \mathbf{f}_F$	Simulated true/healthy/faulty frequency sets, resp.
$\delta\mu$	Fault 1: the relative mean values shift: $\delta\mu = \mu_w - 1$
$\delta\sigma$	Fault 2: the relative difference in the std: $\delta\sigma = \sigma_w/\sigma_v$
γ	The threshold for comparable frequencies
$iter$	The iteration number
$std(\mathcal{N}_s)$	The standard deviation of $\mathcal{N}_s$
$DDIF(std(\mathcal{N}_s))$	The double differentiation of the $std(\mathcal{N}_s)$ curve

6.3.1 Natural Frequency and Mode Shape

We will first briefly introduce natural frequency and mode shape that are used in FTED.

Every structure has tendency to oscillate with much larger amplitude at some frequencies than others. These frequencies are called natural frequencies. When a structure is vibrating under one of its natural frequencies, the corresponding vibration pattern it exhibits is called a mode shape for this natural frequency. For example, we deploy a total of m sensor nodes on a structure and identify a total of p vibration patterns from the measurement of these sensors. The corresponding natural frequency set and mode shapes are denoted respectively as

$$\mathbf{f} = \{f^1, f^2, ..f^p\}, \boldsymbol{\Psi} = [\boldsymbol{\Psi}^1, \boldsymbol{\Psi}^2, \ldots, \boldsymbol{\Psi}^p]$$

where f^k ($k = 1 \ldots, p$) is the kth natural frequency, $\boldsymbol{\Psi}^k (k = 1, \ldots, p)$ is the mode shape corresponding to f^k. $\boldsymbol{\Psi}^k$ is a m-by-1 vector, with each element corresponding to the vibration amplitude at the ith sensor in this vibration pattern.

Both natural frequency and mode shape are internal vibration characteristics of structure and are different for different structures. However, these two characteristics have different usage in the FTED. Natural frequencies are global parameters of a structure which means that, using sensor nodes deployed on different location of a structure, the same set of natural frequencies can be obtained [4]. Another intriguing property of natural frequency is that once damage occurs, all deployed sensors, no matter near or far from the damage location, will still have the same natural frequency set (although it will be different from which obtained when the structure is healthy). This indicates that *when the natural frequencies from a node are significantly different from others, the possible cause should be data fault instead of structural damage.* Considering this property, *natural frequency is taken as the feature for faulty node detection.* However, unless can be identify with high accuracy, natural frequency is not a very sensitive indicator to detect structural damage [4] and does not contain any spatial information for damage localization.

Mode shape Ψ^k has an element corresponding to each sensor and thus contains spatial information. Mode shape and its derivatives, mode shape curvature have been proven to be very sensitive features structural damage [13]. *Therefore, in FTED, mode shape is used to detect structural damage.*

6.3.2 Cluster-Based Natural Frequency Identification

In FTED, both natural frequency and mode shape are identified in a distributed way in the sense that deployed sensors are divided into clusters. Clustering must meet the following constraints: (1) Sensor nodes in each cluster belong to the same substructure. (2) Sensor nodes in each cluster are within the single hop communication range to its cluster head (CH), and (3) All the clusters in the same substructure are connected together through the overlapping nodes.

The first constraint is for faulty node detection. In FTED, natural frequencies of sensors in the same cluster are compared to detect faulty nodes. Although theoretically, natural frequencies should be the same for all the sensor nodes across the structure, the spatial redundancy only preserves well in sensor nodes in the same substructure due to the structural nonlinearity. For example, in a suspension bridge, the natural frequencies extracted from the sensor nodes deployed on cables, piers, or spans can be significantly different.

The last two constraints are for energy-efficient identification of natural frequency and mode shape. Under these constraints, we need to minimize the amount of wireless data transmissions required for identifying natural frequencies and mode shapes. The justification of these two constraints and the objective function is similar to the ones described in [9, 10]. To solve this clustering problem, the basic idea is to first partition the sensor network according to the substructures they belong, and using some greedy heuristic shown in [9, 10] to control the cluster size while still satisfying the last two constraints. The details of clustering are omitted for brevity.

After clustering, the natural frequencies of each sensor node in a cluster are calculated. To achieve this, we first calculate the cross spectral density (CSD) between data from each sensor, relative to the data from a reference node. The reference node can be randomly chosen and is selected as the CH in this chapter. The CSDs of each sensor in a cluster is calculated in a distributed way as in [10]. After calculating the CSD, each sensor node implements the peak-picking (PP) technique [1] to extract its natural frequencies. In the PP, each node picks a number largest peaks from its CSD and the location of each peak corresponds to a natural frequency. A typical CSD and the natural frequencies using the PP method are illustrated in Fig. 6.2, where the extracted natural frequency set is 6, 37, 99, 194, 317, 470.

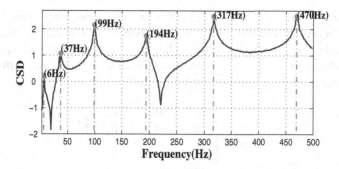

Fig. 6.2 A typical CSD function and the natural frequencies extracted using the PP method

6.3.3 I-FUND

The identified frequency sets in each cluster are transmitted to the CH, where they are compared with each other to detect faulty nodes. Intuitively, sensor nodes which identify statistically different natural frequency sets from others will be labeled as faulty. However, we should note that **the natural frequencies identified in each sensor node is a vector**. Furthermore, two natural frequencies can be compared **only when they belong to the same vibration pattern of structure**. However, some sensor nodes, although they are healthy, may miss identifying some true natural frequencies or erroneously obtain several pseudo ones due to the noise. For example, assume the first four natural frequencies of a structure is $\mathbf{f}_{true} = \{1, 4, 5, 10\}$, where $1, 4, 5, 10$ correspond to the first four vibration patterns of the structure, respectively. However, it is quite possible that the identified natural frequency sets from two healthy sensor nodes are: $\mathbf{f}_1 = \{1.1, 3.0, 4.1, 9.9\}$, and $\mathbf{f}_2 = \{3.9, 5.1, 10.1, 13\}$. By comparing with $\mathbf{f}_{true}$, $\mathbf{f}_1$ contains a fake frequency 3.0 while misses a true one 5, and $\mathbf{f}_2$ missed frequency 1 while falsely identified 13. If $\mathbf{f}_1$ are directly compared with $\mathbf{f}_2$, false positive alarm will be issued. We call this **the element mismatch problem (EMP)**. However, solving the EMP is not an easy task since $\mathbf{f}_{true}$ is never known a priory.

To compare frequency sets with EMP, we propose I-FUND, which counts the number of comparable elements (which are natural frequencies in SHM) in feature vectors iteratively to identify faulty nodes. Before we describe the method, we will give a formal definition of comparable natural frequencies.

Assume using the PP method, natural frequencies from m sensors are extracted, each sensor having p natural frequencies. These frequencies are arranged in a $m - by - p$ matrix:

$$\mathscr{F} = \begin{bmatrix} \mathbf{f}_1 \\ \mathbf{f}_2 \\ \cdots \\ \mathbf{f}_m \end{bmatrix} = \begin{bmatrix} f_1^1 & f_1^2 & \cdots & f_1^p \\ f_2^1 & f_2^2 & \cdots & f_2^p \\ \vdots & \vdots & \ddots & \vdots \\ f_m^1 & f_m^2 & \cdots & f_m^p \end{bmatrix} \tag{6.1}$$

where f_i^k $(i = 1, \ldots, m, k = 1, \ldots, p)$ is the kth natural frequency extracted using the data from ith sensor. Each row of $\mathscr{F}$ belongs to the natural frequency set identified by a particular node. Analogous to the 'adjacency list,' we define a 'comparability list' for each f_i^k, denoted as $clist_{f_i^k}$, which contains all the comparable frequencies of f_i^k in $\mathscr{F}$. The comparability lists of all the frequencies constitute a matrix:

$$\mathscr{F}_{clist} = \begin{bmatrix} clist_{f_1^1} & clist_{f_1^2} & \cdots & clist_{f_1^p} \\ clist_{f_2^1} & clist_{f_2^2} & \cdots & clist_{f_2^p} \\ \vdots & \vdots & \ddots & \vdots \\ clist_{f_m^1} & clist_{f_m^2} & \cdots & clist_{f_m^p} \end{bmatrix}$$

Given $\mathscr{F}$ and a existing $clist_{f_i^k}$ (which is an empty set initially), a frequency f_j^r must satisfy the following conditions to be added into $clist_{f_i^k}$:

1. $\frac{|f_j^r - f_i^k|}{f_j^r + f_i^k} \leq \gamma$, where γ is a threshold defined by user.
2. $i \neq j$,
3. $\forall f_j^\beta (\beta \neq r), f_j^\beta \notin clist_{f_i^k}$, and
4. $\forall clist_{f_i^l} (l \neq k), f_j^r \notin clist_{f_i^l}$

The first constraint is that the f_j^r should be close enough to f_i^k to be added into $clist_{f_i^k}$. The second constraint is set because the comparability list of a frequency should not contain the frequencies from the same sensor. Third, since we aim to establish a one-to-one mapping between frequencies among different sensors, frequencies of the same sensor cannot be contained in the same list. Lastly, due to the same reason, the same natural frequency cannot be contained in more than one comparability lists of the same node.

After calculating $\mathscr{F}_{clist}$, we calculate a supplementary matrix denoted as $\mathscr{N}_s$. $\mathscr{N}_s$ is calculated by adding the cardinalities of comparability lists of each sensor node and is used to evaluate the comparability of each sensor node.

$$\mathscr{N}_s = [\mathscr{N}_{s1}, \ldots, \mathscr{N}_{sm}] = \left[\sum_{k=1}^{p} \left| clist_{f_1^k} \right|, \ldots, \sum_{k=1}^{p} \left| clist_{f_m^k} \right| \right] \qquad (6.2)$$

where $\mathscr{N}_{si}$ is the **comparability of the ith sensor node**. A significant smaller $\mathscr{N}_{si}$ value in $\mathscr{N}_s$ indicates that *on average, natural frequencies of sensor i have low comparability with others and this sensor is therefore labeled as faulty*. We delete faulty sensor node in an iterative manner. Each time when $\mathscr{N}_s$ is calculated, we calculate its standard deviation, denoted as $std(\mathscr{N}_s)$. Then the sensor with minimum $\mathscr{N}_{si}$ will be deleted. $\mathscr{F}_{clist}$ and $\mathscr{N}_s$ are then updated using the remaining nodes. This procedure iterates until the newly obtained $std(\mathscr{N}_s)$ is smaller than a threshold.

The pseudo code for this faulty node detection method is illustrated in Algorithm 6. For convenience, it is called the iterative faulty node detection method (**I-FUND**). It can be seen that I-FUND can be largely divided into two stages: in the first stage, the comparability matrix $\mathscr{F}_{clist}$ is calculated. Then, at the second stage faulty nodes are deleted iteratively.

Algorithm 6 The I-FUND

Input: A universe of sensors and the corresponding $\mathscr{F}$
Output: a group of faulty nodes
1: % **(1) In the** 1st **stage, we calculate the initial** $\mathscr{F}_{clist}$
2: Pick up the first row of $\mathscr{F}$
3: For each frequency f_i^k in this row, find the closest frequency in $\mathscr{F}$, assume its f_j^r
4: **if** $\dfrac{|f_j^r - f_i^k|}{f_j^r + f_i^k} < \gamma$ **then**
5:    add f_j^r into $clist_{f_i^k}$, and block f_j^r for the use of other frequencies in this row
6: **end if**
7: Repeat until all the frequencies in this row has done this search
8: Release all the blocks set in this row
9: Select the next row in $\mathscr{F}$, repeat the procedures to last row.
10: % **(2) In the** 2nd **stage, faulty nodes are deleted iteratively**
11: From $\mathscr{F}_{clist}$, calculate $\mathscr{N}_s$ according to Eq. 6.2
12: **while** $std(\mathscr{N}_s) >$ a threshold **do**
13:    Delete the sensor node with minimum $\mathscr{N}_{si}$ in $\mathscr{N}_s$
14:    Update $\mathscr{F}_{clist}$ and $\mathscr{N}_s$
15: **end while**

We use an example to illustrate the the procedures of I-FUND in a step by step manner. Assume we have four sensors $s_1 \sim s_4$, each identifying 4 natural frequencies

$$\mathscr{F} = \begin{bmatrix} 1 & 4 & 10 & 50 \\ 3.9 & 11 & 20 & 49 \\ 0.2 & 0.9 & 4 & 51 \\ 4 & 6 & 15 & 80 \end{bmatrix} = \begin{bmatrix} f_1^1 & f_1^2 & f_1^3 & f_1^4 \\ f_2^1 & f_2^2 & f_2^3 & f_2^4 \\ f_3^1 & f_3^2 & f_3^3 & f_3^4 \\ f_4^1 & f_4^2 & f_4^3 & f_4^4 \end{bmatrix}$$

Assume $\gamma = 15\,\%$, $\mathscr{F}_{clist}$ can be obtained as

$$\begin{bmatrix} \{f_3^2\} & \{f_2^1, f_3^3, f_4^1\} & \{f_2^2\} & \{f_2^4, f_3^4\} \\ \{f_1^2, f_3^3, f_4^1\} & \{f_3^3\} & \{f_4^3\} & \{f_1^4, f_3^4\} \\ \{\} & \{f_1^1\} & \{f_1^2, f_2^1, f_4^1\} & \{f_1^4, f_2^4\} \\ \{f_1^2, f_2^1, f_3^3\} & \{\} & \{\} & \{\} \end{bmatrix}$$

$\mathscr{N}_s$ is calculated as $\mathscr{N}_s = [7, 7, 6, 3]$, $std(\mathscr{N}_s) = 1.89$.

From $\mathscr{N}_s$ above, s_4 has a minimum value in $\mathscr{N}_s$ than others and this indicates that s_4 is the most incomparable one with others and is therefore deleted. The updated $\mathscr{N}_s$ becomes: $\mathscr{N}_s = [6, 5, 5]$, $std(\mathscr{N}_s) = 0.58$.

If we require that the iteration will stop when $\mathscr{N}_s < 1$, then the iteration stops here. s_4 is the faulty node identified by the I-FUND. It can be seen that in each iteration, a node with smallest comparability among the remaining ones is deleted, indicating that in I-FUND, *nodes with large fault are always deleted first*.

In the I-FUND, the iteration stops when $std(\mathcal{N}_s)$ is smaller than a pre-defined threshold. However, we will show soon that the correct stop point is critical for the performance of I-FUND while such a pre-defined threshold is hard to be correctly determined. In the remaining of this section, we first analyze the performance of the I-FUND by the false positives and false negatives it generates and then describe ROD, an improved criterion to stop the iteration of the I-FUND.

Fault Model

The frequency set from a healthy node, denoted as $\mathbf{f}^H$, is generated by multiplying a random variable v on the true natural frequency set $\mathbf{f}_{true}$: $\mathbf{f}^H = \mathbf{f}_{true} + v * \mathbf{f}_{true}$, where v is a random variable with zero mean (i.e., $\mu_v = 0$) but non-zero standard deviation σ_v. The multiplicative noise instead of additive one is used are justified through real experiment on the PP method. In the similar manner, the natural frequency set of a faulty node, denoted as $\mathbf{f}^D$, is generated by multiplying a random variable w on $\mathbf{f}_{true}$. $\mathbf{f}^D = \mathbf{f}_{true} + w * \mathbf{f}_{true}$. The mean and standard deviation of w are denoted as μ_w and σ_w.

Fault on a sensor node can cause its natural frequencies to be both *inaccurate* and *imprecise*. More formally, the former and the latter can be characterized as

$$\delta\mu = E\left[\frac{|\mathbf{f}^D - \mathbf{f}_{true}|}{\mathbf{f}_{true}}\right] \tag{6.3}$$

and

$$\delta\sigma = std\left[\frac{\mathbf{f}^D - \mathbf{f}_{true}}{\mathbf{f}_{true}}\right] / std\left[\frac{\mathbf{f}^H - \mathbf{f}_{true}}{\mathbf{f}_{true}}\right] - 1, \tag{6.4}$$

respectively, where $E[.]$ represents the mean and $std[.]$ stands for standard deviation.

Obviously, for a faulty node, either its $\delta\mu > 0$, or its $\delta\sigma > 0$, or both of the two conditions are satisfied. It can be seen that $\delta\mu$ and $\delta\sigma$ can be represented in the form of μ_w and σ_w respectively: $\delta\mu = \mu_w$, $\delta\sigma = \sigma_w/\sigma_v - 1$.

To analyze the performance of I-FUND, we generate natural frequency set for 1000 sensor nodes and among them, 800 are healthy. The true frequency set $\mathbf{f}_{true} = [1, 4, 7, 10, 13, 16, 19, 22, 25, 28]$. For healthy nodes, we let $\sigma_v = 0.03$ and for faulty nodes, we let $\delta\mu = 0.06$ while keep $\delta\sigma = 0$. Figure 6.3a shows the comparability of each node (i.e., $\mathcal{N}_{si}$ in Eq. 6.2) before any faulty node is deleted. Without losing any generality, we put the 200 faulty nodes at the end of the 800 healthy ones. Correspondingly, sensor nodes at the left side and the right side of the dashed vertical line in Fig. 6.3a are the healthy ones and faulty nodes, respectively. According to the I-FUND, the one with the minimum $\mathcal{N}_{si}$ will be labeled as faulty and is illustrated as the hollow red dot in Fig. 6.3a.

Figure 6.3b–d shows the $\mathcal{N}_{si}$ at some certain number of iterations, denoted as *iter*, (i.e., *iter* = 50, *iter* = 150, and *iter* = 250, respectively). The nodes that have been labeled as faulty are shown as hollow red dots. It can be seen that with the increase of iteration, more and more faulty nodes are detected (i.e., transformed from '*' to 'o').

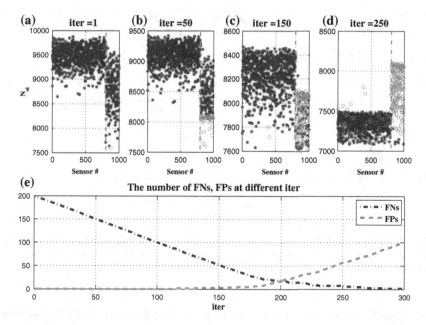

Fig. 6.3 $\mathcal{N}_{si}$ at different iterations when $\delta\mu = 0.06$ **a** Iteration $= 1$, **b** Iteration $= 50$, **c** Iteration $= 150$, **d** Iteration $= 250$. **e** The number of FPs and FNs at different iterations

However, when $iter = 50, 150$, and 200, there still exist some faulty nodes yet to be detected and some healthy nodes erroneously labeled as faulty. The former are called **'false negatives (FNs)'** and the latter are called **'false positives (FPs)'**. Obviously, the number of FNs and FNs are **non-increasing** and **non-decreasing** function w.r.t. *iter*, respectively. As an example, Fig. 6.3e shows the number of FNs and FPs at different *iter*. It can be seen that with the increase of *iter*, the number of FNs decreases and the number of FPs starts to develop after some iterations. *When iter reaches to the number of faulty nodes, denoted as m^F ($m^F = 200$ in this example), the FN curve and the FP curve always intersect with each other.*

To control the number of FPs and FNs, it is critical to find a correct point for I-FUND to stop the iteration. Although the optimal stop point depends on different application requirements on FNs and FNs, to stop the iteration at the intersection of FP and FN curves (i.e., when $iter = m^F$ in Fig. 6.3) is generally a good choice since at this point, the number of FPs and FNs are balanced to some extent. Furthermore, we will show later that in some conditions, $iter = m^F$ can also be the best stop point since at this point, both the number of FNs and FPs are zero. Unfortunately, m^F is not available a priori. We hence have to observe, from the updated $\mathcal{N}_{si}$ at each iteration to see when the iteration should be stopped.

Instead of using a pre-defined threshold for $std(\mathcal{N}_{si})$, we can stop the iteration by observing the changes on $std(\mathcal{N}_s)$. As an example, Fig. 6.4a illustrates the FPs, FNs, and the corresponding $std(\mathcal{N}_s)$ at different *iter* when $\delta\mu = 0.15$. Firstly, we should notice from the upper figure in Fig. 6.4a that in this condition, the optimal

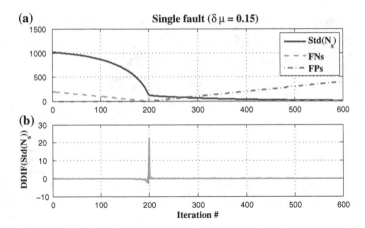

Fig. 6.4 a Using I-FUND on single type of data fault $\delta\mu = 0.15$. *Upper* The FPs, FNs and the $std(\mathcal{N}_s)$ w.r.t. *iter*, *Lower* The $DDIF(std(\mathcal{N}_s))$ w.r.t. *iter*. **b** Using I-FUND on multiple types of faults $\delta\mu = 0.075$ and $\delta\mu = 0.15$

stop point $iter = m^F = 200$ since at this point, both the number of FPs and FNs are zero. Secondly, by observing $std(\mathcal{N}_s)$, we can see that when *iter* is approaching to $m^F = 200$, the corresponding $std(\mathcal{N}_s)$ decreases rather rapidly but this trend suddenly changes after $iter > m^F$. The rationale behind this change of decreasing speed of $std(\mathcal{N}_s)$ at $iter = m^F$ is that deleting a faulty node can cause a much higher decrease in the $std(\mathcal{N}_s)$ than deleting a healthy one. This change before and after $iter = m^F$ in the slope of $std(\mathcal{N}_s)$ can be more clearly observed if the curve of $std(\mathcal{N}_s)$ is double differentiated. The lower figure of Fig. 6.4a illustrates the $DDIF(std(\mathcal{N}_s))$, the double differentiation of $std(\mathcal{N}_s)$. From the clear peak in $DDIF(std(\mathcal{N}_s))$, the optimal *iter* can be obtained.

Although in Fig. 6.4a, the correct stop point is obtained by locating the maximum peak in the $DDIF(std(\mathcal{N}_s))$, in many conditions, particularly when the deployed sensor nodes have different types and levels of faults, deleting nodes with different faults can cause the variation in the descending process of $std(\mathcal{N}_s)$. This small variation in $std(\mathcal{N}_s)$ can generate some large peaks in $DDIF(std(\mathcal{N}_s))$, leading to erroneous stop points. This is demonstrated in Fig. 6.4b, where among 200 faulty nodes, 100 nodes have fault $\delta\mu = 0.075$, and the remaining half have fault $\delta\mu = 0.15$. We can roughly observe two changes of slope in the curve of $std(\mathcal{N}_s)$ and they are approximately located at $iter = 100$ and $iter = 200$, respectively. Correspondingly, $DDIF(std(\mathcal{N}_s))$ has two peaks at these two locations. If the location of the maximum peak in the $DDIF(std(\mathcal{N}_s))$ is used, an erroneous stop point would have been obtained.

To address this problem, an alternative approach is adopted. We first need to make *iter* large enough to guarantee that when iteration stops, most of the faulty nodes have been deleted. This can also be achieved by setting the threshold for $std(\mathcal{N}_s)$ as a very small value. Then *from right to left*, we examine the curve of $DDIF(std(\mathcal{N}_s))$ until we found the first 'statistically large value' in $DDIF(std(\mathcal{N}_s))$. For convenience, we call this method '*reversely outlier detection*' (**ROD**). **The rationale behind the ROD is as follows**: by examining from right to left, we essentially start from a group of healthy nodes, and continue to add more nodes one by one into the group. In case when the variation in frequencies from healthy nodes is relatively small (i.e., small δv), adding healthy frequencies in $\mathscr{F}$ will not cause rapid increase in $std(\mathscr{N}_s)$. Therefore, the slope of $std(\mathscr{N}_s)$ curve at $iter = m^F + 1$ is small. However, when the first faulty node is added into the group, the increase in the $std(\mathscr{N}_s)$ is significant and the slope of $std(\mathscr{N}_s)$ at $iter = m^F$ is large. This change of slope can cause a high value in $DDIF(std(\mathscr{N}_s))$ at this point.

To determine whether a $DDIF(std(\mathscr{N}_s))$ point is statistically significant, statistical process control techniques such as '3-sigma' rule [12] can be applied: we first establish an upper control limit (UCL) using a group of $DDIF(std(\mathscr{N}_s))$ points when $iter$ is very large (i.e., from a group of healthy nodes). Then from right to left, we search each $DDIF(std(\mathscr{N}_s))$ point and see if it exceeds the UCL. This process stops when we meet the first outlier. As an example, Fig. 6.4b shows how the correct point $iter = 200$ is found using the ROD. A thorough evaluation of I-FUND will be given in Sect. 6.4.1.

6.3.4   Identification of Mode Shape

After faulty sensor nodes have been detected by I-FUND, we calculate the mode shapes in each cluster on the remaining sensor nodes, and then mode shapes of clusters in the same substructure are stitched together to identify structural damage.

Identification of mode shapes also relies on the CSD curve but a 'common' natural frequency needs to be first identified before a mode shape can be identified. On the beam structure shown in Fig. 6.5a, a number of sensor nodes are deployed and within which, nodes a, b and c and their CSD curves are highlighted. Applying the PP method on the CSD curves can identify a number of natural frequencies for each sensor. For a cluster of sensors, if a frequency is identified in all the sensors, then there exists a mode shape accordingly. And the mode shape value on each sensor is its CSD amplitude at this frequency point. Figure 6.5b shows three mode shapes as dashed curves. They correspond to natural frequencies f1, f2 and f3, respectively. However, due to the noise, an exact 'common' frequency such as f1/f2/f3 may not exist. In practice, we find a number of special frequency sets called 'universally comparable frequency' (UCF) set from $\mathscr{F}_{clist}$ and the mode shape value on each sensor is its CSD amplitude at the frequency in this UCF set.

We use the example shown in Sect. 6.3.3 to illustrate how to find out the UCF sets. After s_4 is deleted, the $\mathscr{F}_{clist}$ is updated as:

$$\mathscr{F}_{clist} = \begin{bmatrix} \{f_3^2\} & \{f_2^1, f_3^3\} & \{f_2^2, f_3^3\} & \{f_2^3\} \\ \{f_1^2, f_3^3\} & \{f_1^3, f_3^4\} & \{\} & \{\} \\ \{\} & \{f_1^1\} & \{f_1^2, f_2^1\} & \{f_1^3, f_2^2\} \end{bmatrix}$$

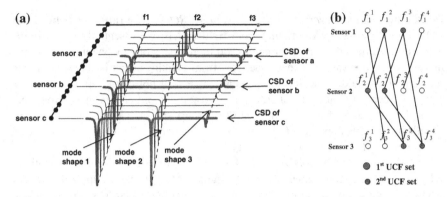

Fig. 6.5 a How mode shapes are identified from the CSDs. **b** The $\mathscr{F}_{clist}$ after s_4 is deleted and the UCF sets

The $\mathscr{F}_{clist}$ can be observed more clearly from the comparability graph shown in Fig. 6.7b, where comparable frequencies are connected with each other. The UCF sets can be obtained from the comparability graph by **looking for the loops which cover all the sensor nodes**. To achieve this, we need only pick one row in the comparability graph and look each element in this row where it is contained in a loop which covers all the sensor nodes. Figure 6.7b also shows the UCF sets, where frequencies in the same UCF sets are marked with the same colors.

After all the clusters in a substructure have identified their mode shapes, these mode shapes are assembled together through the overlapping nodes.

6.4 Simulation

6.4.1 The Performance of I-FUND with ROD Through Simulation

In this section, we will give an extensive evaluation about the performance of the I-FUND with ROD under different conditions. We carried out a total of 4 tests and the parameter settings in each test are shown in Table 6.2.

Table 6.2 The parameter settings in each test

Para.	$\delta\mu$	$\delta\sigma$	δm	σ_v
Test 1	$5\sim 15\%$	0	200/800	0.03
Test 2	0	$0.2\sim 8.2$	200/800	0.03
Test 3	$7, 10\%$	0	$50/950\sim 500/500$	0.03
Test 4	10%	0	200/800	$0.03\sim 0.1$

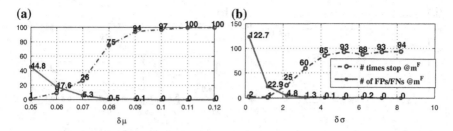

Fig. 6.6 a Test 1: the effect of $\delta\mu$ to the I-FUND, *dashed curve* the total times that I-FUND stop at *iter* = m^F, *solid curve* the average number of FPs/FNs when the I-FUND stops at *iter* = m^F. **b** Test 2: the effect of $\delta\sigma$ to the I-FUND

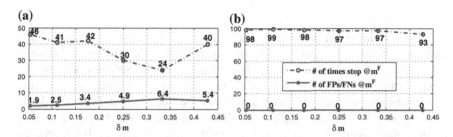

Fig. 6.7 Test 3: the effect of δm to the I-FUND, **a** when $\delta\mu = 0.07$, **b** when $\delta\mu = 0.1$

In the first test, we investigate how the I-FUND work when fault causes 'inaccurate' natural frequencies. Specifically, $\delta\mu$ will be changed from 5 ~ 15% but $\delta\sigma = 0$. Among the 1000 sensor nodes, 200 are faulty with $\delta\mu > 0$. The true frequency set $\mathbf{f}_{true} = [1, 4, 7, 10, 13, 16, 19, 22, 25, 28]$. For each $\delta\mu$ ranging from 5 ~ 15%, we did 100 times simulation and count the times that the I-FUND stops at *iter* = m^F = 200. The results are shown in Fig. 6.6a. It can be observed from the dashed curve that as long as $\delta\mu > 0.1$, I-FUND can stop at *iter* = m^F = 200 with high probability. In the meantime, the average number of FNs/FPs when the I-FUND stops at *iter* = m^F = 200 is shown as the solid curve in the same figure. When $\delta\mu > 0.1$, both the number of FNs and FPs are zero at this point. Therefore, with the setting of this test, I-FUND can work well if the associated fault satisfies $\delta\mu > 0.1$.

In the similar manner, Fig. 6.6b illustrates the performance of the I-FUND when fault causes 'imprecise' natural frequencies where $\delta\sigma$ is changed from 0.2 ~ 8.2. Figure 6.6b shows that the higher the $\delta\sigma$, the easier the algorithm can stop at this point. When $\delta\sigma > 5$, about 90 out of 100 times that the iteration stops at *iter* = 200. And at this point, the number of FPs/FNs is very small.

The third test investigates the effect of the number of faulty nodes. Specifically, the ratio of faulty nodes to the healthy ones, δm, will be changed from 1/999 ~ 500/500. We set fault to be $\delta\mu = 0.07$ and $\delta\mu = 1\%$ and the results are shown in Fig. 6.7a, b, respectively. It is interesting to see that the number of faulty nodes seems to have little effect on the I-FUND, particularly when the fault level is large.

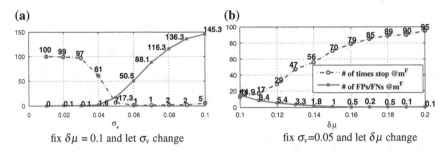

Fig. 6.8 Test 4: the effect of σ_v to the I-FUND. **a** with $\delta\mu = 0.1$, σ_v increase from $0 \sim 0.1$, **b** with $\sigma_v = 0.05$, $\delta\mu$ increase from $0.1 \sim 0.2$

The rationale behind this observation is as follows: In the ROD, we search *from right to left* the first statistically significant point in $DDIF(std(\mathcal{N}_s))$ curve. Obviously, the sharper the change of the slope in $std(\mathcal{N}_s)$ at $iter = m^F$, the better the performance of the method will be. To have a sharp change at $iter = m^F$, only the following two conditions are required: (1) the gap among frequencies from healthy nodes is relatively small, and (2) the gap between healthy frequencies and the frequency of the node with slightest fault is large. The former is able to guarantee that the change of $std(\mathcal{N}_s)$ from $iter = m^F + 2$ to $iter = m^F + 1$ is small, and the latter can make sure that the change of $std(\mathcal{N}_s)$ from $iter = m^F + 1$ to $iter = m^F$ is large. Other factors, including the number of faulty nodes, does not have noticeable effect on this change of the slope at $iter = m^F$.

As part of the illustration of the above observation, we investigate in the last test how the gap among healthy frequencies will affect I-FUND by adjusting σ_v, the standard deviation of the noise added on the healthy frequencies. Firstly, we fix the fault to be $\delta\mu = 0.1$ and let σ_v change from $0.01 \sim 0.1$. From Fig. 6.8a, the effect of noise on healthy frequencies is significant. When $\sigma_v = 0.03$, most of times the algorithm can stop at the right point, and has low FPs/FNs. When $\sigma_v = 0.05$, only 7 out of 100 times that this correct stop point are achieved. On the other hand, Fig. 6.8b shows the results if we fix $\sigma_v = 0.05$ but change $\delta\mu = 0.1 \sim 0.2$. It can be seen that when $\sigma_v = 0.05$, only when the $\delta\mu$ is near 0.2 can the I-FUND work well.

At last, we discuss briefly how γ, the threshold to determine whether two frequencies are comparable, should be set correctly. Too large or too small γ can downgrade the performance of the I-FUND. Generally speaking, γ should be set to a value such that *the gap between the comparabilities of healthy nodes and faulty nodes is maximized*. To have an optimal γ, we generally need to have a training set containing frequencies from both healthy and faulty nodes.

To summarize, given an appropriate γ, the I-FUND can work well where there is a relatively large gap in the comparability of healthy and faulty nodes. This gap is determined by (1) the level of fault (i.e., $\delta\mu$, $\delta\sigma$), and (2) the noise on frequencies from healthy node (i.e., σ_v).

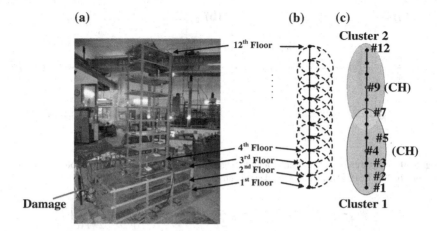

Fig. 6.9 **a** Test structure, **b** network topology, **c** clustering results

6.5 Experiment on a Lab Structure

The effectiveness of the FTED is tested through a real experiment. To address the generally high requirements of SHM application, we designed a particular type of wireless sensor nodes called SHM mote. SHM mote uses Imote2 as its computation and communication unit and contains a sensory board which includes on-board 3-axis accelerometer LIS344ALH and interface for external sensors. The testing structure has 12 floors, and the SHM motes are deployed on each floor to monitor the structure's horizontal accelerations under hammer strike (see Fig. 6.9a). Although the deployed SHM motes can form a complete network in general condition, we use the topology of the network shown in Fig. 6.9b to generate a multi-hop network. The SHM motes run modified TinyOS and are configured to sample the accelerometers at frequency of 256 Hz. Structural damage in this test is generated by releasing a support ring on the third floor (See Fig. 6.9a). Meanwhile, sensors which should be tightly attached on the 5th and 8th floors are released to generate faulty readings on these two nodes. We aim to detect the structural damage in the presence of these two faulty nodes.

This lab structure is relatively simple and the deployed 12 SHM motes can be regarded as in the same substructure. Using the clustering method, two clusters are generated and illustrated in Fig. 6.9c. This clustering is implemented offline in a laptop computer before we do the test. CHs in these two clusters are #4 and #9. In each SHM mote, a natural frequency set is extracted and send back to the corresponding CH. The upper figure in Fig. 6.11 illustrates the identified natural frequency sets and two UCF sets in the first cluster. The lower one shows the $std(\mathcal{N}_s)$, $DDIF(std(\mathcal{N}_s))$ when node is deleted one by one in the I-FUND. From right to left, the first $DDIF(std(\mathcal{N}_s))$ has large peak which exceeds the UCL limits determined by the last 3 sensor nodes. Thus sensor #5 is labeled as faulty. Another faulty sensor node #8 is also detected in another cluster (Fig. 6.10).

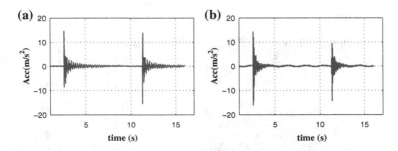

Fig. 6.10 Measurement from healthy and damaged sensors. **a** Data from SHM Mote #4 (healthy sensor). **b** Data from SHM Mote #5 (released sensor)

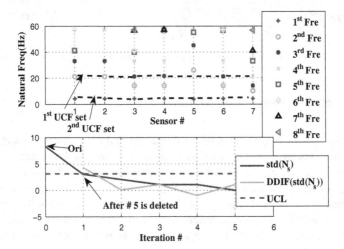

Fig. 6.11 *Upper* identified natural freqs and UCF sets, *Lower* std($\mathcal{N}_s$) after each iteration

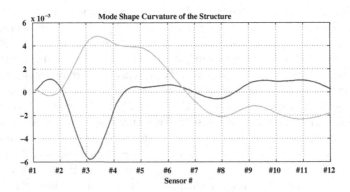

Fig. 6.12 Mode shape curvature of two vibration patterns

After faulty sensor nodes have been detected and isolated, mode shapes of each cluster are estimated based on the two UCF sets shown in Fig. 6.11. In this test, mode shapes corresponding to the two UCF sets are extracted in both clusters and then sent back to the gateway, where the mode shapes at the healthy sensor nodes are first interpolated and then double differentiated to obtain mode shape curvatures. Figure 6.12 illustrates the mode shape curvatures. By examining the curvatures, damage is detected at #3, which matches with the correct damage location.

6.6   Conclusion

In this chapter, we proposed FTED, a scheme to detect complex events in the presence of faulty readings. In FTED, features for faulty node detection are first extracted and based on which, we propose I-FUND with ROD to iteratively detect faulty nodes. The features for event detection are then identified from remaining ones. We believe this architecture, along with the I-FUND with ROD, can be applied to many applications of WSNs besides SHM, where detection of event requires data level collaboration and sampled data at each sensor is a dynamic sequence.

References

1. J.S. Bendat, A.G. Piersol, *Engineering Applications of Correlation and Spectral Analysis* (Wiley-Interscience, New York, 1980), p. 1
2. T. Clouqueur, K.K. Saluja, P. Ramanathan, Fault tolerance in collaborative sensor networks for target detection. IEEE Trans. Comput. **53**(3), 320–333 (2004)
3. M. Ding et al., Localized fault-tolerant event boundary detection in sensor networks, in *INFOCOM*, vol. 2 (2005), pp. 902–913
4. S.W. Doebling, Damage identification and health monitoring of structural and mechanical systems from changes in their vibration characteristics: a literature review. Technical report, Los Alamos National Lab (1996)
5. M.I. Friswell, D.J. Inman, Sensor validation for smart structures. J. Intell. Mater. Syst. Struct. **10**(12), 973–982 (1999)
6. S. Guo, Z. Zhong, T. He, Find: faulty node detection for wireless sensor networks, in *ACM SenSys* (2009), pp, 253–266
7. G. Kerschen, P. De Boe, J.C. Golinval, K. Worden, Sensor validation using principal component analysis. Smart Mater. Struct. **14**, 36 (2005)
8. B. Krishnamachari, S. Iyengar, Distributed Bayesian algorithms for fault-tolerant event region detection in wireless sensor networks. IEEE Trans. Comput. **53**(3), 241–250 (2004)
9. X. Liu, J. Cao, M. Bhuiyan, S. Lai, H. Wu, G. Wang, Fault tolerant wsn-based structural health monitoring, in *IEEE/IFIP 41st International Conference on Dependable Systems and Networks (DSN)*. IEEE (2011), pp. 37–48
10. X. Liu, J. Cao, S. Lai, C. Yang, H. Wu, Y.L. Xu, Energy efficient clustering for wsn-based structural health monitoring, in *Proceedings IEEE INFOCOM*. IEEE (2011), pp. 2768–2776
11. X. Luo, M. Dong, Y. Huang, On distributed fault-tolerant detection in wireless sensor networks. IEEE Trans. Comput. **55**(1), 58–70 (2006)
12. J.S. Oakland, *Statistical Process Control* (Wiley, New York, 1986)

Chapter 7
Conclusions

In this book, we investigate a particular application of wireless sensor networks: structural health monitoring. In this book, we first give a review on existing WSN-based SHM systems, and then introduce our hardware design and software design of WSN-based platform called SenetSHM. Afterward, we address three important aspects that we believe are important when designing a WSN-based SHM system: how to realize event-triggered wakeup, how to implement centralized SHM algorithms within a WSN, and how to address the fault tolerance problem in a WSN-based SHM system. We believe the methods and techniques proposed in this book can serve as a guideline for more applications like SHM.

© The Author(s) 2016
J. Cao and X. Liu, *Wireless Sensor Networks for Structural Health Monitoring*,
SpringerBriefs in Electrical and Computer Engineering,
DOI 10.1007/978-3-319-29034-8_7

Printed in the United States
By Bookmasters